Satellite Information Classification and Interpretation

Edited by Rustam B. Rustamov

Published in London, United Kingdom

IntechOpen

Supporting open minds since 2005

Satellite Information Classification and Interpretation
http://dx.doi.org/10.5772/intechopen.77202
Edited by Rustam B. Rustamov

Contributors
Upendra Rajput, Olga Maltseva, Igor Ivanov, Vladimir Sotskii, Alexandr Tertyshnikov, Gennadii Zhbankov, Juan Manuel Nuñez, Sandra Medina, Gerardo Ávila, Jorge Montejano, Vladimir Lukin, Alexander Zemliachenko, Sergey Krivenko, Benoit Vozel, Kacem Chehdi, Humberto Alves Barbosa, T. V. Lakshmi Kumar, Manoj Kumar Thakur, Franklin Javier Paredes, John R. Mecikalski, Eric Hamsen, Carlos DaCamara, Renata Libonati, Miguel Pinto, Alexandra Hurduc, Jose Navarro-Pedreño, Ignacio Melendez-Pastor, Ignacio Gómez, Gema Marco Dos Santos, Rustam B. Rustamov

First published in London, United Kingdom, 2019 by IntechOpen
IntechOpen is the global imprint of INTECHOPEN LIMITED, registered in England and Wales, registration number: 11086078, The Shard, 25th floor, 32 London Bridge Street
London, SE19SG - United Kingdom
Printed in Croatia

British Library Cataloguing-in-Publication Data
A catalogue record for this book is available from the British Library

Additional hard and PDF copies can be obtained from orders@intechopen.com

Satellite Information Classification and Interpretation
Edited by Rustam B. Rustamov
p. cm.
Print ISBN 978-1-83880-566-1
Online ISBN 978-1-83880-567-8
eBook (PDF) ISBN 978-1-83880-793-1

For EU product safety concerns:
IN TECH d.o.o., Prolaz Marije Krucifikse Kozulić 3, 51000 Rijeka, Croatia, info@intechopen.com or visit our website at intechopen.com.

Meet the editor

Dr. Rustam B. Rustamov is a freelance expert on space science and technology. He was in charge of Azerbaijan National Aerospace Agency activities as Acting Director General. Dr. Rustamov specializes mainly in space instrumentation and remote sensing and GIS technology. He was invited to the work at the European Space Agency within the Framework of the United Nations Program on Space Applications at the European Space Research and Technology Center, The Netherlands. Rustam B. Rustamov is a member of the Cambridge Scholars Publishing International Editorial Advisory Board, Astronomy & Astrophysics, London, UK. He is an author of 15 books, published under major publishers in Europe and United States, and he has written and published over 120 scientific papers.

Contents

Preface

The first stage of satellite information classification and interpretation is data collection using remote sensing. Space technology includes both satellite and aerial remote sensing applications. In general, these applications operate at different spectra of the electromagnetic radiation, as energy from the Sun reaches Earth's surface and is again reflected, transmitted, or absorbed by the objects that collected it via satellite sensors or recorded it in the satellite's memory.

The next stage of classifying and interpreting data is making a finer interpretation of spectra for identifying Earth's features. Significant advances in sensor technology stemmed from subdividing the spectral ranges of electromagnetic radiation into several bands, thus allowing sensors across these bands to form multispectral images, opening up opportunities for bringing into sharp relief Earth's features at high accuracy and increased segments of processing. In general, there are three different types of data products: black and white photograph or panchromatic image (single band), normal color, and false color composite (multichannel). Single band images display as grayscale, but a combination of three bands at a time generates color composite images.

Interpretation of satellite information may be visual or digital, or it may integrate both modes, containing the line of the processes of detection, identification, description, and assessment of the detected object.

Visible imagery is due to radiation in the electromagnetic wave range of 0.4–0.7 μ; it is available during daylight hours and when atmospheric transparence is good. Some satellites can sense low-intensity visible light at night, but these data are not routinely used by operational meteorologists. In general, visible imagery is black and white. White is used for the brightest and most reflective energy received by the sensor, whereas black displays the least reflective values. Low brightness is associated with oceans, lakes, and the background of Earth; medium brightness values come from land, including forests and deserts. Clouds produce high brightness, displayed in white or light gray.

Atmospheric windows are generally used for signal detection from Earth. One such is near infrared, covering wavelengths of 0.75–1.4 μm. Infrared (or thermal infrared, IR) imagery is derived from terrestrial radiation emitted by Earth, cloud tops, and the atmosphere in the range of 10–12 μ. This portion of the spectrum is available 24 hours a day and does not depend on atmospheric conditions. IR values are a measure of the temperature of the emitting surface, with some modifications due to absorption and reemission as the radiation passes through the atmosphere.

A complication of interpreting IR images is due to their lower resolution (relative to wavelengths of the visible spectrum). Healthy vegetation reflects infrared radiation much more strongly than it reflects green energy, appearing very bright in the image. A simple example is the light tone appearance given by vegetation species and the dark tone given by water. Particularly, in thermal infrared images, brightness represents the warmest temperature and darkness the coolest.

In radar imagery, smooth surfaces reflect highly, whereas areas blocked from radar signals appear dark. Bridges and cities appear very bright, while calm water, pavement, and dry lake beds appear very dark.

The basic elements of radar image interpretation are the following:

- Shape—The external form, outline, or configuration of the object revealing its natural features;
- Size—A property that depends on the scale and resolution of the image/photo. Smaller features are easily apparent in a large-scale image/photo;
- Pattern—The spatial arrangement of an object into distinctive recurring forms, as is illustrated in the pattern of a road or railway line;
- Shadow—This indicates the outline of an object and its length, which is useful in measuring the height of an object;
- Tone—It is a feature that refers to the color or relative brightness of an object; tonal variation is due to the reflection, emittance, transmission, or absorption character of an object, which may vary from one object to another and may change with reference to different bands;
- Resolution—It includes spectral and spatial resolutions and depends on the photographic/imaging device used, namely, whether camera or sensor. Spectral resolution helps identify features seen in specific spectral bands. Advances in sensor development have helped bring about high-resolution imagery, which assists planners and professionals to produce large-scale maps and improve their planning and monitoring. High spatial resolution imagery/photographs are useful in identifying small objects. A map's scale and the classification level of information it can provide are mainly determined by the spatial resolution of the image/photograph.

Rustam B. Rustamov
Khazar University,
Azerbaijan

Section 1

Aerospace Image Processing

Chapter 1

Introductory Chapter: Aerospace Information Classification

Rustam B. Rustamov

1. Introduction

1.1 Classification of space images

Classification of the aerospace image is the process of creating an environment systemizing the image pixel values into meaningful segments of the Earth features. There are a number of methods and facilities for classification of aerospace images. Aerospace image classification methods can be formed into three categories:

- Automatic
- Manual
- Hybrid

The methods indicated above can be used independently depending on task and available technological maintenance. There is no doubt that each of the methods has advantages and disadvantages. In general, automatic image classification method is the most preferable one used in spatial data classification [1].

It is obvious that aerospace image classification takes a vital place from the first stage of line of image processing up to producing final electronic or hard production. **Figure 1** demonstrates a flowchart of the processes as the required actions in the aerospace image classification.

Based on the above approach, we can describe standard approach as:

- Spatial data
- Extract information for an application
- Visual and digital image interpretation
- Field survey
- Integration spatial data into field survey
- Thematic map creation
- Decision-making

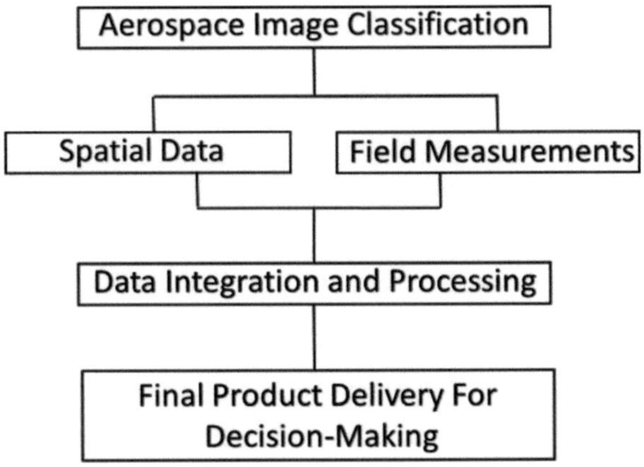

Figure 1.
Aerospace image processing classification flowchart.

1.2 Aerospace images

During the work with aerospace images, first of all, the spectral wavelength of shooting is important for researchers which defines:

- Biogeophysical characteristics of objects transferred by images and technology of achieving the image

which depend on

- Visualization, radiometric, and geometrical properties of images

These two characteristics represent a basis of the classification of aerospace images considering possibilities of their processing.

Spectral wavelength of shooting defines the first, fundamental, level of this classification considering the reflective and radiating characteristics of the objects reproduced in the image. From this point of view, three main types of aerospace images are defined:

- In visible, near and middle to the infrared wavelength
- In a thermal infrared wavelength
- In radio frequency wavelength

As the general understanding of the classification is the procedure of related to the object belonging to one of Q-classes. The relation to the object comes and defines on the basis of presence on an object of some features. Classifications are result the objects divided into the classes [2, 3].

Classification process of aerospace images in general provides the following stages:

- Identification of the main distinctive features (characteristics) reflecting from different objects with different classes
- Creation of spatial features

- Definition (calculation) at the studied object of feature(s) on which bases expected to classified of feature(s)
- Decision-making about object belonging to one of the classes with application of the decisive rule

1.3 Remote sensing data processing

Classification of remote sensing data is the process which is used for receiving aerospace images of the maps of the land surface and any other information sources which depends on task solution. Subjects of the map can be contained in a variety of segments from enough general categories, such as:

- Soil
- Vegetation
- Surface water

and up to thinner structures, for example:

- Various types of soils
- Species of vegetation
- Depth of reservoirs

The subject of classification in such tasks is the image pixel, the features of which conducts definition of the classes such as brightness of this pixel. It is necessary to indicate that in general, aerospace images consist of several layers corresponding to different spectral channels. From this point of view, brightness is the vector value. Coordinates of the vector define the provision/location of the pixel in the spatial of feature. The value of the field in the same point of coordinate is defined by the quantity of the pixels which is located into this point.

1.4 Aerospace image processing

The computer algorithm analysis realizing various procedures of classification is subdivided into two types:

- Autonomous classification
- Classification with training

Computer processing of the aerospace images presented in the digital form opens new technical capabilities for image processing. Special software packages used for the preparation of this subject, such as ERDAS Imagine, allow to display the images for monitoring, improve quality of the images (e.g., to remove influence of an atmospheric impact), synthesize color images, carry out the automated processing, and obtain quantitative data (coordinates, distances, the areas, etc.). Results of computer processing form a basis for creation of maps which can be written in a digital form or printed on the paper [4–6].

Receive the aerospace images when shooting the scanning systems from different aero- and satellite systems; for instance, the French SPOT or the American Landsat is used for processing. By means of high-precision scanners, they can be transferred to a digital format and images.

The digital images consist of the segments, pixels forming a grid of lines, and columns. Each pixel has the coordinates and characterized by brightness which is designated in conventional units. The value of brightness is connected with ability of terrestrial objects to reflect solar radiation. From this aspect it shows how significantly performed in the images of differences in brightness of objects and depends the result of processing that is:

- Objects brightness characteristics
- Use of multichannel images
- Synthesis of the color image
- The automated processing with use of appropriate software application
- Measurements according to the images

Aerospace image classification has a variety of applications such as remote sensing, image and data storage for transmission in business applications, etc. It is important to use advances of spatial data classification in wide areas, particularly in Earth studies.

Author details

Rustam B. Rustamov
eiLink Research and Development of Khazar University, Azerbaijan

*Address all correspondence to: r_rustamov@hotmail.com

References

[1] Siddaraju K, Nagaraju D, Bhanuprakash HM, Shivaswamy HM, Balasubramanian A. Morphometric evaluation and sub basin analysis in Hanur Watershed, Kollegal Taluk, Chamarajanagar District, Karnataka, India, using Remote Sensing and GIS Techniques. International Journal of Advanced Remote Sensing and GIS. 2017;**6**(1):2178-2191. ISSN: 2320-0243

[2] Colomina I, Molina P. Unmanned aerial systems for photogrammetry and remote sensing: A review. ISPRS Journal of Photogrammetry and Remote Sensing. 2014;**92**:79-97

[3] Sugg ZP, Finke T, Goodrich DC, Moran MS, Yool SR. Mapping impervious surfaces using object-oriented classification in a semiarid urban region. Photogrammetric Engineering and Remote Sensing. 2014;**80**(4):343-352

[4] Gupta K, Shweta S, Singh A, Aryan I, Kuliyal S, Aniruddha D, et al. Geospatial Techniques for Urban Regeneration, Heritage Conservation and Planning. International Journal of SPA. 2017;**12** (Article in Press)

[5] Shaohui S, Carl S. Aerial 3D building detection and modeling from airborne LiDAR point clouds. IEEE Journal of Selected Topics in Applied Earth Observations and Remote Sensing. 2013;**6**(3):1440-1449

[6] Rikimaru A, Miyatake S. Development of forest canopy density mapping and monitoring model using indices of vegetation, bare soil and shadow. In: Proceedings of the 18th Asian Conference on Remote Sensing (ACRS); Kuala Lumpur, Malaysia. International Journal of Advancement in Remote Sensing, GIS and Geography. **5**(117):17

Section 2

Spatial Frequency Representation

Chapter 2

Pan-sharpening Using Spatial-frequency Method

Upendra Kumar

Abstract

Over the years, researchers have formulated various techniques for pan sharpening that attempt to minimize the spectral distortion, i.e., retain the maximum spectral fidelity of the MS images. On the other hand, if the use of the PAN-sharpened image is just to produce maps for better visual interpretation, then the spectral distortion is not of much concern, as the goal is to produce images with high contrast. To solve the color distortion problem, methods based on spatial frequency domain have been introduced and have demonstrated superior performance in terms of producing high spectral fidelity pan-sharpened images over spatial-scale methods.

Keywords: pan sharpening, spatial scale, spatial frequency analysis, discrete wavelet transform, non–subsampled contourlet transform, pseudo-Wigner distribution, urban planning

1. Introduction

Earth resource satellites provide data covering different parts of the electromagnetic spectrum at different spatial, spectral, and temporal resolutions. To utilize these different types of image data effectively, a number of pan-sharpening techniques have been developed [1].

Further, in order to benchmark different image fusion techniques, image quality metrics have been used. There are two types of metrics used to evaluate image quality, namely, subjective (qualitative) and objective (quantitative). The objective of this chapter is to discuss the methodology of some of the prevalent existing techniques, as well as the mathematical representation of some of the standard existing evaluation indicators.

2. Pan-sharpening techniques

Pan sharpening is also known as image fusion, image integration, and multisensor data fusion. Over the years, a large number of pan-sharpening techniques have been developed and have placed into different categorizes. In this study, multiscale transform (MST)-based techniques have been discussed.

2.1 Multiscale transform-based pan-sharpening techniques

In recent years, multiscale transform (MST)-based pan-sharpening techniques have received a lot of attention, since they preserve the spectral fidelity in the pan-sharpened images. Further, it is more suitable for information representation, interpretation, and analysis [2, 3].

Many variations of the multiscale transform-based techniques exist, such as discrete wavelet transform (DWT), stationary wavelet transform (SWT), curvelet transform (CVT), contourlet transform (CT), and Non–subsampled contourlet transform (NSCT) [4]. The next subsections give a descriptive overview and methodology of MST-based pan-sharpening techniques which are selected for this study.

2.1.1 Discrete wavelet transform (DWT)

Before discussing about discrete wavelet transform, first of all, it would be appropriate to discuss in general regarding Fourier transform (FT).

Fourier transform (FT) was first invented by French mathematician and physicist Jean Baptiste Joseph Fourier in 1822. Fourier stated that any periodical function can be represented as a sum of sine and cosine of different frequencies, each multiplied by a different coefficient [5, 6]. Fourier transform converts a signal from the time-amplitude domain to the frequency-amplitude domain. Images are considered as 2-D discrete functions. To use Fourier transform to analyze images, discrete Fourier transform (DFT) is used. FT is a reversible transform, which means the original signal can be recovered through the inverse discrete Fourier transform (IDFT) [7, 8].

However, FT has a drawback, i.e., it does not provide the information about the time at which the particular frequency exists in the signal. Fourier transform only captures the different frequencies in a signal and cannot detect when those frequencies occurred. To overcome this drawback, wavelet transform (WT) was introduced. Wavelet transform (WT) can be more useful than Fourier transform, since it is based on functions that are localized in both space and frequency/scale [9]. Wavelet transform brings a multiresolution framework. With this setting, the signal can be decomposed into components that collect the information at a specified scale, i.e., different frequencies are analyzed with different resolutions [2–6]. The WT has numerous applications in remote sensing such as image registration, spatial and spectral fusion, feature extraction, speckle reduction, texture classification, and crop phenology detection [7].

Wavelet transform can be broadly classified into two main groups, i.e., continuous wavelet transform (CWT) and discrete wavelet transform (DWT). Since CWT is continuous, as a result, there are an infinite number of scale and translation parameters which leads to an infinite number of possible wavelet functions. To overcome the shortcoming of CWT, DWT was introduced.

In the DWT algorithm, an image can be analyzed by passing it through an analysis filter bank followed by decimation operation. The analysis filter bank consists of low pass and high pass filter at each decomposition stage. When a signal passes through these filters, it splits in to two signals. The low pass filter, which corresponds to an averaging operation, extracts the coarse (average) information of the signal. The high pass filter, which corresponds to a differencing operation, extracts the detail information of the signal such as edges, points, and lines. The output of the filtering operation is then decimated by two, i.e., a 2-D transform is accomplished by performing two separate one-dimensional transform [9–12]. First of all, the image is filtered along the row and decimated by two, and it is then followed by filtering the subimage along the column and decimated by two.

This operation splits the image into four bands namely one approximation band, which contains coarse information and three detail bands, horizontal, vertical, and diagonal, respectively, which contain information about the salient features of the image such as edges, points, and lines [5, 8]. A J-level decomposition can be performed resulting in $(3j + 1)$ different frequency bands. At each level of decomposition, the image is split into high and low frequency components; the low-frequency components can be further decomposed until the desired resolution is reached [13–15]. The pan-sharpening procedure for the pan sharpening of panchromatic (PAN) and multispectral (MS) images using DWT has been explained in Section 3.1 (**Figure 1**).

2.1.2 Stationary wavelet transform (SWT)

It is observed that discrete wavelet transform (DWT) is not a shift-invariant transform. Therefore, in order to get rid of this problem, stationary wavelet transform (SWT)-based fusion technique, an extension of DWT scheme, also known as "à trous" algorithm, has been introduced [10, 11]. In the "à trous" algorithm, the downsampling step is suppressed and instead the filter is upsampled by inserting zeros between the filter coefficients (**Figure 2**).

In the SWT algorithm, it uses a two-dimensional filter derived from the scaling function. This produces two images, of which one is an approximation image while the other is a detailed image called the wavelet plane. A wavelet plane represents the horizontal, vertical, and diagonal detail between 2^j and 2^{j-1} resolution and is

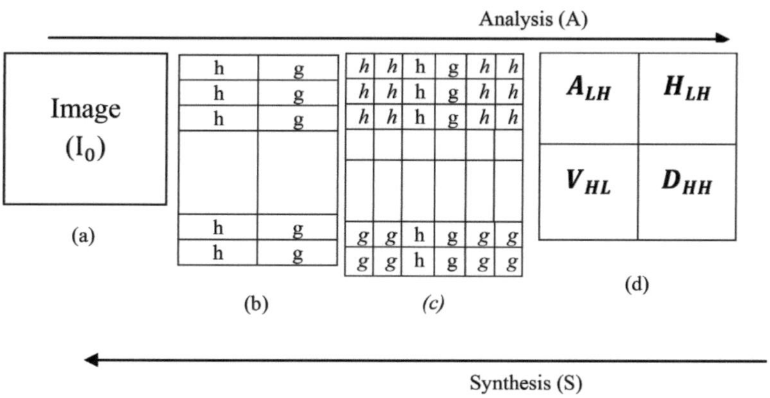

Figure 1.
Decomposition of an image using DWT.

$$\frac{1}{256}\begin{bmatrix}1 & 4 & 6 & 4 & 1\\ 4 & 16 & 24 & 16 & 4\\ 6 & 24 & 36 & 24 & 6\\ 4 & 16 & 24 & 16 & 4\\ 1 & 4 & 6 & 4 & 1\end{bmatrix} \longrightarrow \boxed{\uparrow 2} \longrightarrow \frac{1}{256}\begin{bmatrix}1 & 0 & 4 & 0 & 6 & 0 & 4 & 0 & 1\\ 0 & 0 & 0 & 0 & 0 & 0 & 0 & 0 & 0\\ 4 & 0 & 16 & 0 & 24 & 0 & 16 & 0 & 4\\ 0 & 0 & 0 & 0 & 0 & 0 & 0 & 0 & 0\\ 6 & 0 & 24 & 0 & 36 & 0 & 24 & 0 & 6\\ 0 & 0 & 0 & 0 & 0 & 0 & 0 & 0 & 0\\ 4 & 0 & 16 & 0 & 24 & 0 & 16 & 0 & 4\\ 0 & 0 & 0 & 0 & 0 & 0 & 0 & 0 & 0\\ 1 & 0 & 4 & 0 & 6 & 0 & 4 & 0 & 1\end{bmatrix}$$

$$\mathbf{F_1 = F_j[n]} \qquad\qquad \mathbf{F_2 = F_{j+1}[n]}$$

Figure 2.
Structure of "à trous" filters.

computed as the difference between two consecutive approximations I_{l-1} and I_l levels. All the approximation images obtained, by applying this decomposition, have the same number of columns and rows as the original image, since filters at each level are upsampled by inserting zeros between the filter coefficients and make the size of the image same [16–19].

This is a consequence of the fact that the "à trous" algorithm is a nonorthogonal, redundant oversampled transform [19–21]. The "à trous" decomposition process is shown in **Figure 2**.

The procedure for the pan sharpening of PAN and MS images using SWT can be summarized as follows (**Figure 3**):

i. To generate new panchromatic images, match histograms of PAN image to their corresponding MS image.

ii. Perform the second-level wavelet transform only on the modified PAN image.

iii. The resulting wavelet planes of PAN are added directly to each MS images.

The SWT eliminates the shift sensitivity problem at the cost of an overcomplete signal representation. However, it does not resolve the problem of feature orientation. In addition, the discrete wavelet transform (DWT), and stationary wavelet transform (SWT), cannot capture curves and edges of images well. Wavelets perform well only at representing point singularities, i.e., appropriate to represent linear edges, since they ignore the geometric properties of structures and do not exploit the regularity of edges.

For curved edges, the accuracy of edge localization in the wavelet transform is low. So, there is a need for an alternative approach, which has the potential or capability to detect, represent, and process high-dimensional data. In order to solve this problem, multiscale geometric analysis has been further investigated. As a result, Candès and Donoho [22] have proposed the concept of curvelet transform (CVT).

Further, in order to solve the problem of curvelet transform, which is first developed in continuous domain and then does discretization of images or signals of interest, Yang et al. [23] and Do and Vetterli [24] presented a flexible multiresolution, local, and directional image expansion using contour segments, named contourlet transform. However, due to the downsampling and upsampling, the CT lacks shift invariance and thus results in ringing artifacts [16]. To overcome the weakness of wavelets, curvelets, and contourlets, Cunha et al. [25] proposed

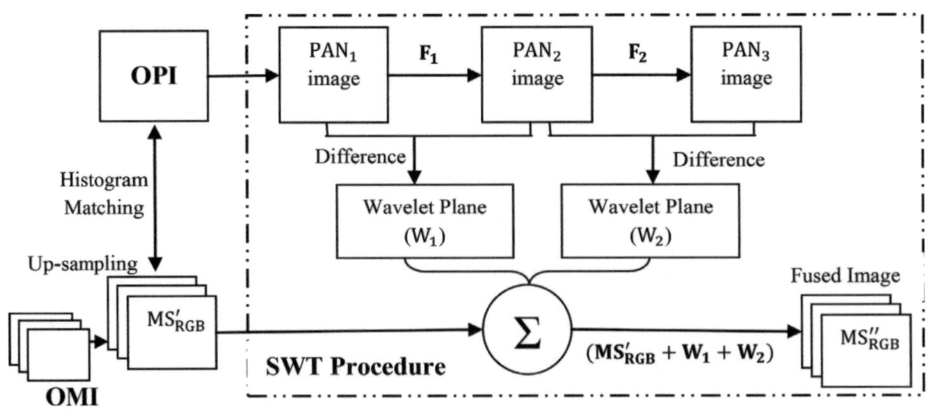

Figure 3.
Methodology adopted for SWT-based pan-sharpening.

non–subsampled contourlet transform (NSCT), based on non–subsampled pyramid decomposition (NSPD) and non–subsampled filter bank (NSFB).

2.1.3 Non–subsampled contourlet transform (NSCT) technique

In order to reduce the frequency aliasing of contourlets and enhance directional selectivity and shift invariance, Holschneider and Tchamitchian [17] proposed non–subsampled contourlet transform. This is based on the non–subsampled pyramid filter banks (NSPFBs) and the non–subsampled directional filter banks (NSDFBs) structure. The former provides multiscale decomposition using two-channel non–subsampled 2-D filter banks, while the later provides directional decomposition, i.e., it is used to split band pass subbands in each scale into different directions [25, 26].

As a result, NSCT is shift invariant and leads to have better frequency selectivity and regularity than CT [25–28]. The scheme of NSCT structure is shown in **Figure 4 (a)**. The NSCT structure classifies two-dimensional frequency domain into wedge-shaped directional subband as shown in **Figure 4(b)**.

In order to provide more practical and flexible solution to the existing problem as stated above, there is a need for an improved or a new fusion technique, which is superior among all the existing pan-sharpening techniques. A new pan-sharpening technique should ideally possess properties of shift invariance, directionality, low computational complexity, and low computational time, applicable to real-time image processing tool, and is also efficient in capturing intrinsic geometrical structures of the natural image along the smooth contours. Moreover, it should perform efficiently under all categories of datasets, such as very high, high, and medium resolution satellite datasets. A spatial frequency-based technique should ideally possess properties, such as shift invariance, directionality, low computational complexity, and low computational time, applicable to real-time image processing tool, and is also efficient in capturing intrinsic geometrical structures of the natural image along the smooth contours [27, 28]. Thus, in order to resolve the existing problems, pan-sharpening method based on joint spatial frequency domain such as pseudo-Wigner distribution has been introduced.

3. Spatial-frequency based pan-sharpening technique

Analysis of non-stationary 2-D signals (image) is a challenging job, as their spectral properties change with time. Such signals cannot be analyzed well by pure spatial domain and frequency domain representations. The joint spatial frequency domain-based image analysis methods, such as Wigner Ville distribution (WVD)

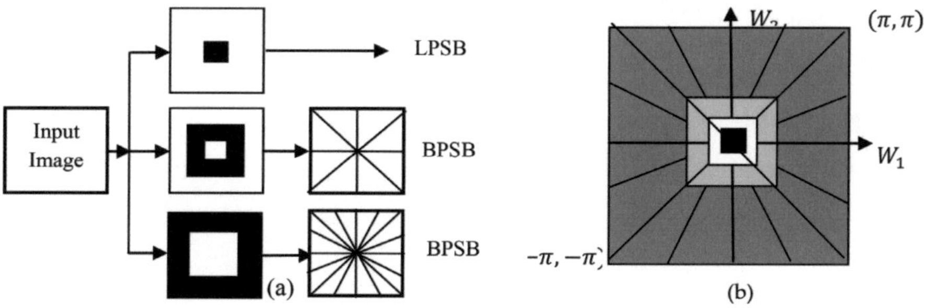

Figure 4.
Two level NSCT decomposition. (a) NSFB structure that implements the NSCT and (b) the corresponding frequency partition.

and pseudo-Wigner distribution (PWD), have been proven to be a powerful tool for analyzing, understanding, and detection of spatial frequency characteristics of non-stationary images in a more comprehensive manner.

The use of Wigner Ville distribution for image processing was first suggested by [18]. It was shown that WVD is a very efficient and powerful tool for capturing the essential non-stationary image structures [29] and appears as a new promising method for the characterization of local spectral properties of images. The Wigner Ville distribution has many interesting properties related to translation, modulation, scaling, convolution, and localization in spatial frequency space, real-valued function and contains phase information, which motivates its use in the field of image analysis applications. Since WVD suffers with the serious problem of interference that makes the interpretation impossible, thus to resolve the limitation of WVD, pseudo-Wigner distribution (PWD) was introduced.

3.1 Pseudo-Wigner Distribution (PWD) technique

Spatial frequency information of a non-stationary image can be effectively extracted with one of the well-known spatial frequency technique known as pseudo-Wigner distribution (PWD). PWD is ideally suited for representing a nonstationary image in the spatial frequency domain and is carried out by adapting the fast Fourier transform (FFT) algorithm. The significant properties of PWD motivate its use in the field of image processing, especially for the fusion of satellite images [30, 31]. These properties are as follows:

i. PWD provides a pixel-wise analysis, efficient and powerful tool for capturing the essential nonstationary image structures, as well as for the characterization of local spectral properties of images, which is indispensable for image fusion.

ii. PWD is shift-invariant technique. Shift-invariant property is necessary for a high-quality and effective image fusion. In the absence of shift invariance, artifacts, such as aliasing effect, loss of linear continuity in spatial features, become prevalent in the resulting fused image (**Figure 5**).

iii. Multidirection, i.e., the window can be tilted in any direction to obtain a directional distribution.

iv. Computation time for PWD is generally small.

With reference to **Table 1**, pseudo-Wigner distribution (PWD) overcomes the shortcomings of the traditional Fourier-based methods, discrete wavelet transform (DWT), stationary wavelet transform (SWT), curvelet transform (CT), contourlet

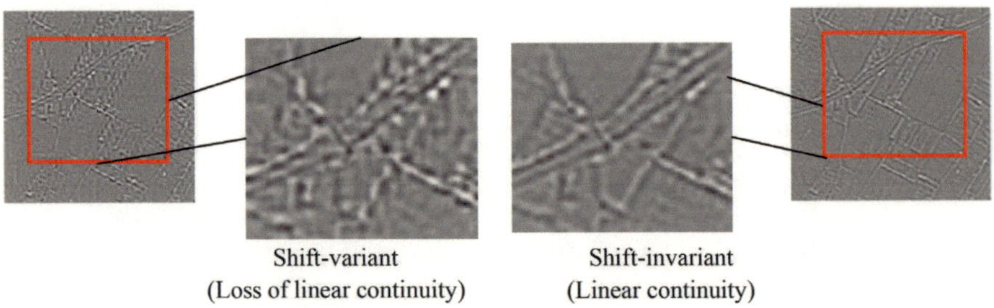

Figure 5.
Concept of shift variant and shift invariant.

Method	Shortcoming
DWT	Poor directionality, lack of shift invariance
SWT	Limited directional selectivity
NSCT	Time-consuming, blocking artifacts

Table 1.
Shortcomings of existing pan-sharpening methods.

transform (CT), and non–subsampled contourlet transform (NSCT). Consequently, it is not based on a multiscale decomposition procedure as wavelets and contourlets are. Further, one of the most challenging applications that comes across by the remote sensing experts is to fuse MS and PAN images collected from different or same satellite sensor with each other to achieve a pan-sharpened image, without introducing artifacts or inconsistencies; otherwise it may damage the quality of the fused image.

Thus, the goal of pan-sharpening is to produce pan-sharpened images with the highest spectral fidelity possible, as the importance of such images in various applications, ranging from land use/land cover classification to road extraction. Therefore, preserving the spectral information of the original MS images in the pan-sharpened images is of great importance [31–33]. Therefore, an attempt to utilize the concept of pseudo-Wigner distribution (PWD) for the pan-sharpening of high-resolution PAN image with a low-resolution MS image has been introduced.

3.1.1 Mathematical background of pseudo-Wigner distribution

Let us consider an arbitrary 1-D discrete function $v(n)$. The PWD of a given array $v(n)$ of N pixels is given by Eq. (1).

$$W(n,m) = 2\sum_{k=-\frac{N}{2}}^{\frac{N}{2}-1} v(n+k)v^*(n-k) \times \exp\left(-2i\left(\frac{2\pi k}{N}\right)m\right) \tag{1}$$

where n and m represent the spatial and frequency discrete variables, respectively, and k is a shifting parameter. Eq. (1) can be interpreted as the discrete Fourier transform (DFT) of the product $v(n+k)v{*}(n-k)$. Here, v^* indicates the complex conjugate of 1-D sequence, v. $W(n,m)$ is a matrix where every row represents the pixel-wise PWD of the pixel at position n. Further, $v[n]$ is a 1-D sequence of data from the image, containing the gray values of N pixels, aligned in the desired direction. By scanning the image with a 1-D window of N pixels, i.e., shifting the window to all possible positions over the full image, the full pixel-wise PWD of the image is produced. The window can be tilted in any direction to obtain a directional distribution [34, 35]. Further, the reasons for selecting short 1-D window for PWD analysis are as follows:

i. It greatly decreases the computational cost.

ii. It allows to obtain a pixel-wise spectral analysis of the data.

The general pan-sharpening procedure adopted for the pan sharpening of PAN and MS images using DWT, NSCT, and PWD [35] techniques can be summarized as follows (**Figure 6**):

i. Coregister both the source images and resample the multispectral image to make its pixel size equal to that of the PAN, in order to avoid the problem of misregistration.

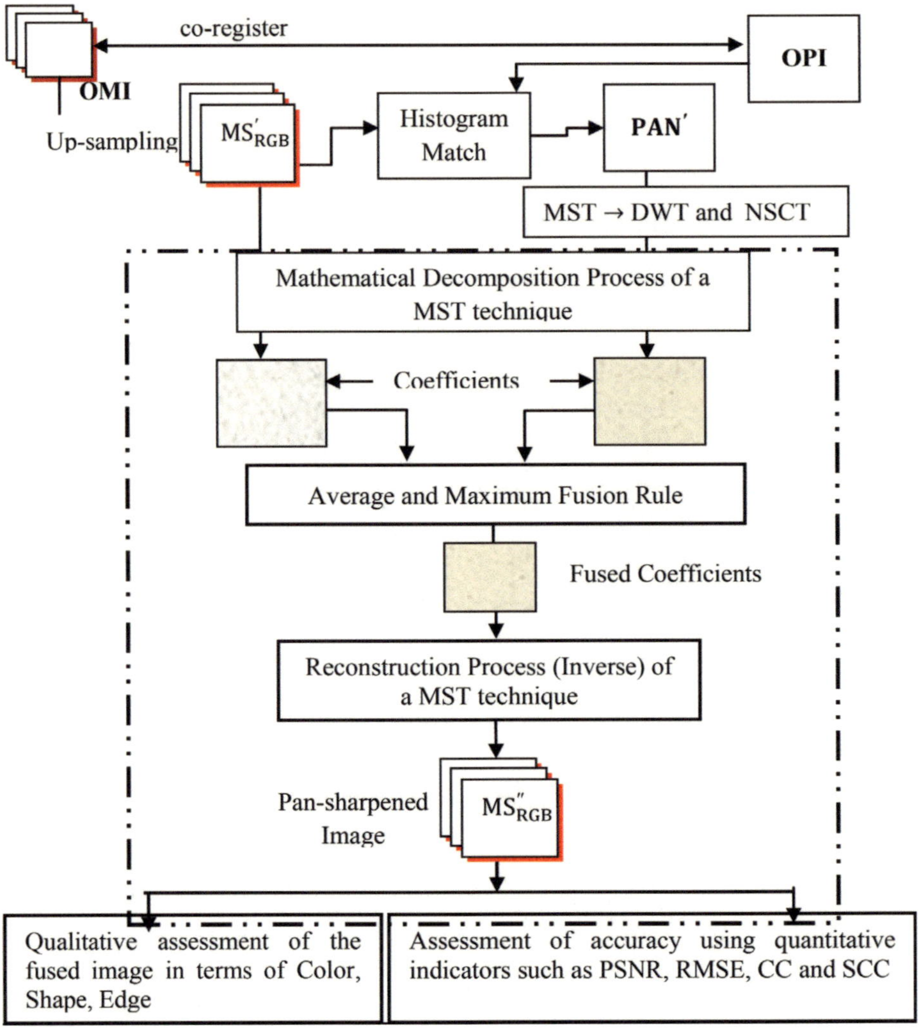

Figure 6.
General methodology adopted for DWT-, NSCT-, and PWD-based pan-sharpening.

ii. Apply DWT/NSCT/PWD to all input coregistered images, one by one, to get their respective coefficients according to the mathematical decomposition procedure related to each one of the techniques, along with upsampling and histogram matching.

iii. The obtained coefficients generated in step (i) from the different input images, i.e., MS and histogram-matched PAN image, are combined according to defined fusion rules to get the fused coefficients.

iv. The fused coefficients are subject to an inverse DWT/NSCT/PWD to construct the fused image. As a result, a new MS image with higher spatial resolution is obtained.

As a result, a new multispectral image with higher spatial resolution is obtained. This process is repeated for each individual MS and PAN band pair. Finally, all the new fused bands are concatenated to form a new fused multispectral image.

Grade	Absolute measure	Relative measure
1	Excellent	The best in group
2	Good	Better than the average in group
3	Average	Average level in group
4	Poor	Lower than the average level
5	Very poor	The lowest in the group

Table 2.
Assessment of image quality by qualitative method.

It may be noted that each MST technique (DWT, NSCT, and PWD) has its unique mathematical properties, which leads to different image decomposition procedure of an image.

3.2 Comparative assessment of various pan-sharpening techniques

Pan-sharpening techniques, belonging to color, statistical, and multiscale transform-based techniques, have been evaluated in terms of certain parameters, such as spectral distortion, shift invariance, directionality, and computational complexity. Comparative assessment of various pan-sharpening techniques has been shown in **Table 2**.

4. Fusion rules

There are various fusion rules to combine the fused coefficients. Let $W_A^P(x,y)$ and $W_B^{MS}(x,y)$ denote the coefficients for higher spatial resolution PAN image and for the lower spatial resolution MS image, and $W^F(x,y)$ denotes the coefficient of the fused image. Using these notations, following fusion rules can be summarized as follows:

i. Average fusion rule

The average fusion rule takes the average of the coefficients of the $W_A^P(x,y)$, PAN, and $W_B^{MS}(x,y)$, MS images, which is given by Eq. (2).

$$W^F(x,y) = \left(W_A^{PAN}(x,y) + W_B^{MS}(x,y)\right)/2 \tag{2}$$

ii. Maximum fusion rule

The maximum fusion rule compares the coefficients from the $W_A^{PAN}(x,y)$, PAN, and $W_B^{MS}(x,y)$, MS images, and picks the larger magnitudes as the fused coefficients, which is given by Eq. (3).

$$W^F(x,y) = \begin{cases} W_A^{PAN}(x,y), & if\ \left|W_A^P(x,y)\right| > \left|W_B^{MS}(x,y)\right| \\ W_B^{MS}(x,y), & if\ \left|W_A^P(x,y)\right| \le \left|W_B^{MS}(x,y)\right| \end{cases} \tag{3}$$

Here, both the fusion rules are chosen as the basic fusion rule throughout this study, which are explained by Eqs. (2) and (3).

5. Assessment of accuracy for pan-sharpening techniques

Pan-sharpening algorithms are designed to produce good-quality pan-sharpened images. A fused image would be considered perfect quality if the spatial detail missing in the MS image is transferred from the panchromatic image without distorting the spectral content of the multispectral image [26]. Unfortunately, this is not possible. There is a trade-off between enhancement of spatial detail and spectral distortion. A fully spatially enhanced fused image would be the panchromatic (PAN) image itself, while an image free of spectral distortion would be the original multispectral (MS) image [36].

The diversity of datasets has contributed to the development of different types of techniques and procedures for the implementation of image fusion. In order to benchmark different pan-sharpening techniques, image quality metrics have been used, i.e., quality metrics are required to evaluate the quality of the fused images [37, 38]. There are two types of metrics used to evaluate image quality:

i. Subjective (qualitative)

ii. Objective (quantitative)

5.1 Qualitative evaluation

Qualitative analysis deals with the visual comparison of the original PAN and MS images with that of the fused image, in terms of spectral and spatial distortion. The evaluation results vary depending on the intensity, sharpness, existence of noisy areas, missing spatial detail, and distortions in the geometry of the objects and display conditions of the image. A number of viewers will be shown the images and asked to judge the image quality. These may also vary from observer to observer, i.e., interpretation of image quality may be influenced or varied by personal preference [39, 40]. Therefore, an exact decision cannot be given. Further, these methods are time-consuming, inconvenient, and expensive.

On the basis of expert/observer personal preference, quality of fused image has been ranked in terms of "Grade," "Absolute Measure," and "Relative Measure" [41], as shown in **Table 2**.

5.2 Quantitative evaluation metrics

It is evident that, in most cases, there is slight difference among fusion results, i.e., quantitative evaluation methods sometimes produce results that cannot be sustained by visual inspection. However, there is no universally accepted metric to objectively evaluate the image fusion results. The generated pan-sharpened images are compared from diverse perspectives of image visualization, coherence, structural similarity, and spectral information content.

The well-known full-reference objective metrics are correlation coefficient (CC), root mean square error, peak signal-to-noise ratio [41]. The reason behind selecting these evaluation indicators is that they measure the statistical, structural similarity, and spectral distortion introduced by the pan-sharpening process. The quantitative metrics that are used in this study, as well as the mathematical representation of these measures, have been discussed below.

5.2.1 Root mean square error

Root mean square error (RMSE) is a frequently used measure of the differences between the fused and the original images. RMSE is a good measure of accuracy [41]. Smaller RMSE value represents a greater accuracy measure and is explained by Eq. (4).

$$RMSE = \sqrt{\sum_{i=1}^{m}\sum_{j=1}^{n}\frac{\left(F(i,j) - R_o(i,j)\right)^2}{c}} \tag{4}$$

where $m \times n$ indicates size of the image and $F(i,j)$ and $R_o(i,j)$ indicate the fused image and the original image, respectively.

5.2.2 Peak signal-to-noise ratio

Peak signal-to-noise ratio (PSNR) indices reveal that the radiometric distortion of the fused image is compared to the original image. PSNR can reflect the quality of reconstruction. The larger value of PSNR indicates less amount of image distortion [41] and is given by Eq. (5).

$$PSNR = 10\ \log\left(\frac{L}{RMSE}\right)^2 \tag{5}$$

where L is related to the radiometric resolution of the sensor; for example, L is 255 for an 8-bit sensor and 2047 for a 16-bit sensor.

5.2.3 Correlation coefficient

The correlation coefficient (CC) of two images is often used to indicate their degree of correlation. If the correlation coefficient of two images approaches one, it indicates that the fused image and original image match perfectly [40, 41]. High value of the correlation shows that the spectral characteristic of the multispectral image has been preserved well. The correlation coefficient is represented by Eq. (6)

$$corr(x,y) = \frac{\sum_{i=1}^{m}\sum_{j=1}^{n}(x(i,j) - \overline{x})(y(i,j) - \overline{y})}{\sqrt{\sum_{i=1}^{m}\sum_{j=1}^{n}(x(i,j) - \overline{x})^2\sum_{i=1}^{m}\sum_{j=1}^{n}(y(i,j) - \overline{y})^2}} \tag{6}$$

where $x(i,j)$ and $y(i,j)$ are the elements of the images x and y, respectively, and $\overline{x}$ and $\overline{y}$ stand for their mean values.

5.2.4 Spatial correlation coefficient

In order to assess the spatial quality of the fused image quantitatively, procedure proposed by [42] has been adopted. This approach is used to measure the amount of edge information from the PAN image, which is transferred into the fused images. The high spatial resolution information missing in the MS image is present in the high frequencies of the PAN image. The pan-sharpening process inserts the higher frequencies from the PAN image into the MS image. Therefore, the CC between the high pass filtered PAN and the fused images would indicate how much spatial information from the PAN image has been incorporated into the MS image. A higher correlation between the two high pass filtered images implies that the spatial

Metric	Ideal value	Error value
Root mean square error (RMSE)	0	>0
Peak signal-to-noise ratio (PSNR)	NA	>1
Correlation coefficient (CC)	1	> − 1 and <1
Spatial correlation coefficient (SCC)	1	> − 1 and <1

Table 3.
The ideal and error value of different quantitative indicators.

information has been retained faithfully. This CC is called the spatial correlation coefficient (SCC). In order to extract the spatial detail of the images to be compared, following Laplacian filter has been used and is represented by Eq. (7).

$$\text{Mask} = \begin{bmatrix} -1 & -1 & -1 \\ -1 & 8 & -1 \\ -1 & -1 & -1 \end{bmatrix} \tag{7}$$

The pan-sharpened image which will best preserve the spectral and structural information of the original low resolution MS image is the one that has satisfied the following conditions (**Table 3**).

6. Summary

This chapter provides the methodology of the proposed approaches for the pan-sharpening of satellite images, along with the discussion of some prevalent existing multisensor pan-sharpening techniques and well-known evaluation indicators.

Author details

Upendra Kumar
Ambalika Institute of Management and Technology, Lucknow, India

*Address all correspondence to: upendra2122@gmail.com

References

[1] Abdikan S, Sanli FB, Sunar F, Ehlers M. A comparative data-fusion analysis of multi-sensor satellite images. International Journal of Digital Earth. 2012;7(8):671-687

[2] Miao Z, Shi W, Samat A, Lisini G, Gamba P. Information fusion for urban road extraction from VHR optical satellite images. IEEE Journal of Selected Topics in Applied Earth Observations and Remote Sensing. 2016;**9**:1817-1829

[3] Alparone L, Baronti S, Aiazzi B, Garzelli A. Spatial methods for multispectral pansharpening: Multiresolution analysis demystified. IEEE Transactions on Geoscience and Remote Sensing. 2016;**54**(5):2563-2576

[4] Khademi G, Ghassemian H. Incorporating an adaptive image prior model into Bayesian fusion of multispectral and panchromatic images. IEEE Geoscience and Remote Sensing Letters. 2018;**15**(6):917-921

[5] Garguet-Duport B, Girel J, Chassery JM, Pautou G. The use of multi resolution analysis and wavelets transform for merging SPOT panchromatic and multispectral image data. Photogrammetric Engineering and Remote Sensing. 1996;**62**(9):1057-1066

[6] Gonzalez RC, Woods RE, Eddins SL. Digital Image Processing Using MATLAB (Vol. 624). Upper Saddle River, New Jersey: Pearson-Prentice-Hall; 2004

[7] Gungor O. Multi-sensor multi-resolution image fusion [Ph.D.]. West Lafayette, Indiana: Purdue University; 2008

[8] Vivone G, Alparone L, Chanussot J, Dalla Mura M, Garzelli A, Licciardi GA, et al. A critical comparison among pansharpening algorithms. IEEE Transactions on Geoscience and Remote Sensing. 2015;**53**(5):2565-2586

[9] Dogra A, Goyal B, Agrawal S. From multi-scale decomposition to non-multi-scale decomposition methods: A comprehensive survey of image fusion techniques and its applications. IEEE Access. 2017;**5**:16040-16067

[10] Mallat S. A theory for multi-resolution signal decomposition: The wavelet representation. IEEE Transactions on Pattern Analysis and Machine Intelligence. 1989;**11**(7):674-693

[11] Polikar R. The Wavelet Tutorial. Theory and Applications of Wavelets. 2008

[12] Ranchin T, Wald L. The wavelet transform for the analysis of remotely sensed images. International Journal of Remote Sensing. 1993;**14**(3):615-619

[13] Roy S, Howlader T, Mahbubur Rahman SM. Image fusion technique using multivariate statistical model for wavelet coefficients. Signal, Image and Video Processing. 2013;7(2):355-365

[14] Vidakovic B, Mueller P. Wavelets for Kids: A Tutorial Introduction. Duke University, Durham, NC: Institute of Statistics and Decision Science, 2007

[15] Pradhan B, Jebur MN, Shafri HZM, Tehrany MS. Data fusion technique using wavelet transform and Taguchi methods for automatic landslide detection from airborne laser scanning data and quickbird satellite imagery. IEEE Transactions on Geoscience and Remote Sensing. 2016;**54**(3):1610-1622

[16] Starck JL, Murtagh F. Image restoration with noise suppression using the wavelet transforms. Astronomy and Astrophysics. 1994;**288**:342-348

[17] Holschneider M, Tchamitchian P. Regularite´ local de la function non-differentiable' the Riemann. In: Lemarie PG, editor. Les Ondelettes en 1989. Berlin, Heidelberg: Springer; 1990. pp. 102-124

[18] González-Audícana M, Otazu X, Fors O. A comparison between Mallat's and the a'trous discrete wavelet transform based algorithms for the fusion of multispectral and panchromatic images. International Journal of Remote Sensing. 2005;**26**(3): 595-614

[19] Jiang Q , Jin X, Lee SJ, Yao S. A novel multi-focus image fusion method based on stationary wavelet transform and local features of fuzzy sets. IEEE Access. 2017;**5**:20286-20302

[20] Medina J, Carrillo I, Upegui E. Spectral and spatial assessment of the TDW wavelet transform decimated and not decimated for the fusion of OrbView-2 satellite images. In: 2018 13th Iberian Conference on Information Systems and Technologies (CISTI). IEEE; 2018

[21] Vetterli M, Kovačević J. Wavelets and Subband Coding. Englewood Cliffs, NJ: Prentice Hall; 1995

[22] Candès EJ, Donoho DL. Curvelet multi resolution representation and scaling laws. In: Proceedings of Eighth SPIE Wavelet Applications in Signal and Image Processing. 2000. p. 4119

[23] Yang Y, Tong S, Huang S, Lin P, Fang Y. A hybrid method for multi-focus image fusion based on fast discrete curvelet transform. IEEE Access. 2017;**5**:14898-14913

[24] Do MN, Vetterli M. The contourlet transform: An efficient directional a multi-resolution image representation. IEEE Transactions on Image Processing. 2005;**14**(12):2091-2106

[25] Cunha AL, Zhou J, Do MN. Non sub sampled contourlet transform: Filter design and applications in de-noising. In: IEEE International Conference on Image Processing. Vol. 1. 2005. pp. 749-752

[26] Yang XH, Jiao LC. Fusion algorithm for remote sensing images based on non sub sampled contourlet transform. Acta Automatica Sinica. 2008;**34**(3):274-281

[27] Ding S, Zhao X, Xu H, Zhu Q , Xue Y. NSCT-PCNN image fusion based on image gradient motivation. IET Computer Vision. 2017;**12**(4):377-383

[28] Shabanzade F, Ghassemian H. Combination of wavelet and contourlet transforms for PET and MRI image fusion. In: Artificial Intelligence and Signal Processing Conference (AISP). IEEE; 2017. pp. 178-183

[29] Claasen TACM, Mecklenbrauker WFG. The Wigner distribution—A tool for time–frequency analysis. Philips Journal of Research. 1980;**35**(3):217-250

[30] Gabarda S, Cristobal G. On the use of a joint spatial-frequency representation for the fusion of multi-focus images. Pattern Recognition Letters. 2005;**26**(16):2572-2578

[31] Rajput UK, Ghosh SK, Kumar A. Multi-sensor fusion of satellite images for urban information extraction using pseudo-Wigner distribution. Journal of Applied Remote Sensing. 2014;**8**(1): 083-668

[32] Gabarda S, Cristóbal G. Blind image quality assessment through anisotropy. Journal Optical of Society America A. 2007;**24**(12):B42-B51

[33] Redondo R, Fischer S, Sroubek F, Cristobal G. 2D Wigner distribution based multi-size windows technique for image fusion. Journal of Visual Communication and Image Representation. 2008;**19**(1):12-19

[34] Redondo R, Sroubek F, Fischer S, Cristóbal G. Multi-focus image fusion using the log-Gabor transform and a multi-size windows technique. Information Fusion. 2009;**10**(2):163-171

[35] Rajput UK, Ghosh SK, Kumar A. Comparison of fusion techniques for very high resolution data for extraction of urban land-cover. Journal of the Indian Society of Remote Sensing. 2015; **45**(4):709-724

[36] Vijayaraj V, Younan N, O'Hara C. Quantitative analysis of pan-sharpened images. Optical Engineering. 2006; **45**(4):46-202

[37] Wald L, Ranchin T, Mangolini M. Fusion of satellite images of different spatial resolutions: Assessing the quality of resulting images. Photogrammetric Engineering and Remote Sensing. 1997; **63**(6):691-699

[38] Chen H, Varshney PK. A human perception inspired quality metric for image fusion based on regional information. Information Fusion. 2007; **8**(2):193-207

[39] Dhore AD, Veena CS. Evaluation of various pan-sharpening methods using image quality metrics. In: 2015 2nd International Conference on Electronics and Communication Systems (ICECS). IEEE; 2015. pp. 871-877

[40] Shi W, Zhu CO, Tian Y, Nichol J. Wavelet-based image fusion and quality assessment. International Journal of Applied Earth Observation and Geo-information. 2005;**6**(3):241-251

[41] Karathanassi V, Kolokousis P, Ioannidou S. A comparison study on fusion methods using evaluation indicators. International Journal of Remote Sensing. 2007;**28**(10):2309-2341

[42] Zhou J, Civco DL, Silander JA. A wavelet transform method to merge land sat TM and SPOT panchromatic data. International Journal of Remote Sensing. 1998;**19**(4):743-757

Chapter 3

Lossy Compression of Remote Sensing Images with Controllable Distortions

Vladimir Lukin, Alexander Zemliachenko, Sergey Krivenko, Benoit Vozel and Kacem Chehdi

Abstract

In this chapter, approaches to provide a desired quality of remote sensing images compressed in a lossy manner are considered. It is shown that, under certain conditions, this can be done automatically and quickly using prediction of coder performance parameters. The main parameters (metrics) are mean square error (MSE) or peak signal-to-noise ratio (PSNR) of introduced losses (distortions) although prediction of other important metrics is also possible. Having such a prediction, it becomes possible to set a quantization step of a coder in a proper manner to provide distortions of a desired level or less without compression/decompression iterations for single-channel image. It is shown that this approach can be also exploited in three-dimensional (3D) compression of multichannel images to produce a larger compression ratio (CR) for the same or less introduced distortions as for component-wise compression of multichannel data. The proposed methods are verified for test and real life images.

Keywords: lossy compression, remote sensing, image processing, performance prediction

1. Introduction

A huge amount of data is provided nowadays by existing remote sensing (RS) sensors, both spaceborne and airborne [1, 2]. Data volume is especially large if images are hyperspectral (i.e., having hundreds sub-band images) and/or high resolution ones. Note that both tendencies (to create and exploit multichannel systems as well as to produce high resolution data) are typical for recent years. Volume of acquired data additionally increases due to more frequent observations of sensed terrains [2]—it has become a usual practice to monitor a territory quite often, e.g., each week.

The obtained RS data have to be transferred, stored and/or disseminated. For each of this operation, data compression can be desirable [1, 3, 4]. Meanwhile, there are several obstacles that can prevent efficient execution of these operations. Concerning data transferring: bandwidth of a communication channel used to transfer data can be limited, time for transferring can be restricted, time and power for compression can be limited as well [1, 3]. The same can relate to data dissemination although the limitations are usually less strict compared to downlink data

IntechOpen

transferring. Memory for RS data storage can be a problem too despite of rapid development of new facilities in recent years [2].

Therefore, it is often desired to compress RS images [4, 5]. As known, there are lossless and lossy image compression techniques [1]. Limits attainable by lossless compression are practically reached [1]. Compression ratio (CR) for the existing methods rarely reaches 5 even for compressing hyperspectral data when inter-band correlation is exploited in full extent [4]. However, larger CR values are required often. Then, lossy compression of acquired RS data has to be applied.

The main peculiarity of lossy compression is that it introduces losses (distortions, degradations) into RS images. Then, it can be useful only under condition that introduced losses do not sufficiently negatively influence the goals the acquired RS data are intended for (terrain classification and/or parameter estimation, specific object detection, etc.). One assumption is that introduced losses have to be of the same level or smaller than degradations due to noise in original data [6]. Therefore, noise characteristics have to be taken into consideration and, thus, they should be known in advance or pre-estimated [7–11]. This also means that it is necessary to be able to control introduced distortions and/or to provide a desired level of losses. Moreover, often this should be done automatically, e.g., in on-board compression [3, 12].

A slightly other assumption is possible if compressed images are subject to visual inspection and analysis. Then, introduced distortions should be such that they do not degrade image visual quality [13]. Then, one has to take into account both specific properties of component images, e.g., variations of their dynamic range [7, 14, 15] and peculiarities of human vision system (HVS).

Finally, one more assumption is that introduced distortions should be such that they do not have (noticeable) negative impact on classification accuracy or performance of other operations of RS data processing at final stages. Note that classification accuracy reduction is connected with metrics characterizing introduced distortions [16].

Thus, introduced distortions should be controlled for all aforementioned strategies. Here by "controlled" we mean several aspects. First, distortions have to be measured or estimated or predicted to ensure that they are not larger than allowed threshold according to a certain metric (criterion) [17, 18]. Second, introduced distortions can be accurately measured only if compression and decompression are already done. Then, if distortion level has to be changed, coder parameters have to be changed and metric calculation has to be done after next iteration of compression/decompression [18]. This is often impractical, especially on-board. Then, it is more reasonable to talk about distortion estimation or prediction without compression and decompression but with approximate providing of a desired quality of compressed data.

Certainly, CR can be important as well. Then, an appropriate compromise has to be provided between CR and introduced losses. Note that CR also depends upon a used coder and a way data redundancy is exploited. In this sense, it is worth incorporating inter-channel correlation inherent for multichannel RS data that can be done in different ways [19–21]. It is possible to apply different transforms [11, 22–24] or to carry out different groupings of component images [11, 25, 26].

Lossy compression of images with taking into account noise type [27] and characteristics has been paid considerable attention [28–30]. Possible existence of optimal operation point (OOP) and its prediction have been claimed and studied [13, 18]. Problems of CR prediction and its providing for coders based on discrete cosine transform (DCT) have been considered [18, 31]. Meanwhile, problems of prediction of compressed image quality and providing a desired quality have not been thoroughly analyzed yet.

In this direction, a certain work has been done. In particular, an approach to quality prediction for wavelet based compression of remote sensing images has been put forward [32]. Prediction of mean square error (MSE) of introduced losses for JPEG has been done [33]. However, control and prediction of metric values for more advanced coders as AGU [34] and ADCT [35] that outperform JPEG considerably [36] were not developed till last 2 years. Since providing of a desired metric value using iterative (multiple) compression/decompression requires sufficient time and resources [36], it was decided to design a new approach without iterations [37]. Later this approach has been further advanced [38–40], mainly for single-component (grayscale) images in 8-bit representation and with taking into account possible presence of noise.

In this chapter, we consider application of the designed approach to RS images including multichannel data and keeping in mind the following: (1) dynamic range of component images in multichannel data varies in wide limits and 16-bit representation is often used for them; (2) in many component images of multichannel (e.g., hyperspectral) data, input peak signal-to-noise ratio (PSNR) is high and noise influence is negligible; (3) there is essential correlation of signal component in neighbor sub-band images of multichannel images. We show that by taking into account these properties, it is possible to carry out efficient compression of multichannel RS data with controllable quality.

2. Peculiarities of RS image lossy compression

To understand the problem of lossy compression, some preliminaries are needed.

First, lossy compression introduces distortions due to which a decompressed image differs from the corresponding original one (subject to compression). These distortions are introduced at the stage of quantization of coefficients of a used orthogonal transform: wavelet, DCT or some other [34, 35, 41]. If DCT serves as the basis of lossy compression, quantization step (QS) or scaling factor (SF) serve as parameter that controls compression (PCC). A larger QS or SF leads, in general, to greater introduced distortions and a larger CR [34, 35] but MSE of introduced losses and attained CR values considerably depend upon complexity of a compressed image and noise presence.

Figure 1 presents three images: noise-free image Frisco of low complexity, the same image corrupted with additive white Gaussian noise with zero mean and variance 100, and noise-free image Airfield of quite high complexity (it contains a lot of edges and fine details).

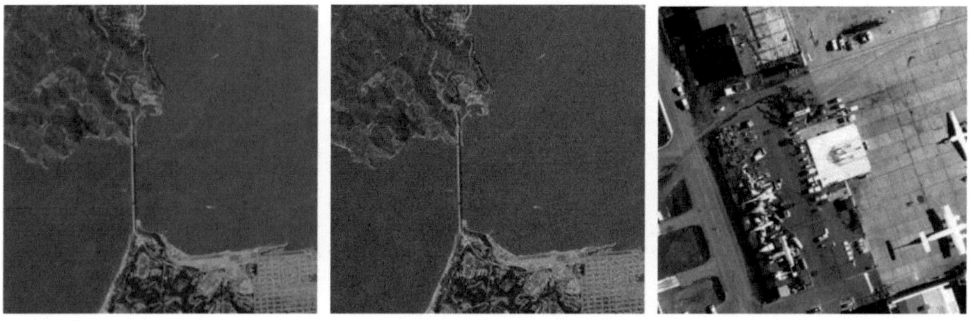

Figure 1.
Noise-free and noisy ($\sigma = 10$) test images Frisco and the test image airfield.

Figure 2 shows dependences of mean square error MSE_{out} between original and compressed images on QS for the case the advanced DCT (ADCT) coder [42] is applied. It is seen well that smaller distortions are introduced if an image is noise-free and has a simpler structure. The values of $MSE_{out}(QS)$ for the same QS can differ by several times and, thus, i.e., QS itself does not determine $MSE_{out}(QS)$.

Dependences CR(QS) for the same images are presented in **Figure 3**. It is seen that the simple structure noise-free image Frisco is compressed in the best way whilst the complex structure image Airfield is compressed with the smallest CR. The reason is that the percentage of DCT coefficients that are assigned zero values after quantization increases if image complexity is lower, noise intensity is less, and QS is larger [31, 43]. Thus, the rate/distortion curve is individual for each particular image and QS has to be adapted to image and noise properties to provide a desired compromise or to satisfy imposed requirements.

We have already mentioned that compression of noisy images has several peculiarities. Suppose that an acquired (noisy) image in a k-th component is image is represented as [8, 10]

$$I_{kij}^{noisy} = I_{kij}^{true} + n_{kij}\left(I_{kij}^{true}\right), i = 1, \ldots, I, j = 1, \ldots, J, k = 1, \ldots, K \qquad (1)$$

where I_{kij}^{noisy} is the ijth sample of the kth component image, n_{kij} is the ijth value of the in the kth component image supposed dependent on I_{kij}^{true} - the true value for the kijth voxel, I and J define the image size, K is the number of components. One can determine input MSE for each component image as

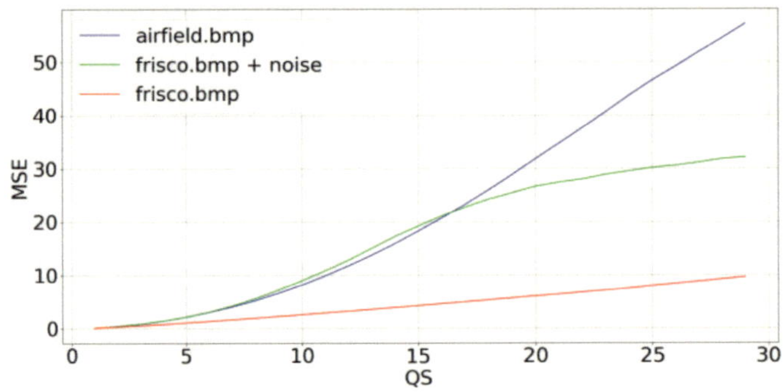

Figure 2.
Dependences MSE vs QS for noise-free and noisy images Frisco and noise-free image airfield.

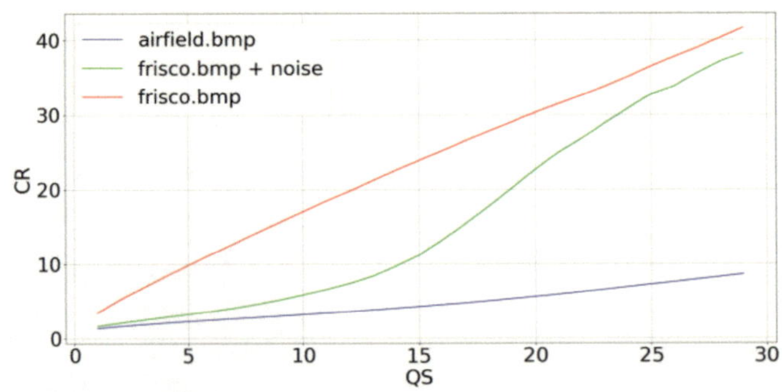

Figure 3.
Dependences CR vs QS for noise-free and noisy images Frisco and noise-free image airfield.

$$MSE_k^{inp} = \sum_{i=1}^{I}\sum_{j=1}^{J}\left(I_{kij}^{noisy} - I_{kij}^{true}\right)^2/(IJ), \quad k = 1, \dots, K \tag{2}$$

and, respectively, input PSNR

$$PSNR_k^{inp} = 10\log_{10}\left(D_k^2/MSE_k^{inp}\right), k = 1, \dots, K, \tag{3}$$

where D_k is image dynamic range assumed individual for each component image ($D_k = I_k^{max} - I_k^{min}$ where I_k^{max} and I_k^{min} are maximal and minimal values in the kth image, respectively).

Earlier analysis [7, 44] has shown that MSE_k^{inp}, k = 1, ..., K and $PSNR_k^{inp}$, k = 1, ..., K in very wide limits for such typical examples of multichannel RS data as images provided by hyperspectral sensors AVIRIS [45] and Hyperion [46]. For more than 80% of component images, input PSNR exceeds 40 dB. This means that, most probably [42], OOPs for these component images do not exist, i.e. $MSE_k^c = \sum_{i=1}^{I}\sum_{j=1}^{J}\left(I_{kij}^c - I_{kij}^{true}\right)^2/(IJ)$ steadily increases if QS becomes larger ($\{I_{kij}^c, i = 1, \dots, I, j = 1, \dots, J, k = 1, \dots, K\}$ denotes compressed image in a k-th channel; OOP exists for a k-th component image if $MSE_k^c(QS)$) has one minimum).

If so, i.e. if quality of the compressed noisy image steadily decreases with QS growth, there should be some reasonable strategy to carry out compression for such an image or a group of images with similar properties. Here it is worth recalling the following. Analysis done in the paper [16] has shown that lossy compression has practically no negative impact on image classification accuracy if the metric PSNR-HVS-M [47] is not less than 42–44 dB.

The metric PSNR-HVS-M ($PSNR - HVS - M_k^c = 10\log_{10}\left(D_k^2/MSE_k^{HVSM}\right)$, k = 1, ..., K, MSE_k^{HVSM} is MSE with taking into consideration specific features of human vision system (HVS)) takes into account two important peculiarities of human vision system: less sensitivity to degradations in high spatial frequencies and masking effect of textures. One can be surprised that visual quality metric has been used in analysis. This can be explained by the fact that the required values of PSNR-HVS-M > 42 dB mean that quality of a compressed image is such that introduced distortions are invisible. According to PSNR, this happens if $PSNR_k^c$ exceeds 35–37 dB [48].

Thus, we need to provide a desired (controlled) quality of compressed images. This should be done quickly (desirably, without iterative compression/decompression), rather accurately, and with producing a large CR. We expect that CR increase can be gained due to grouping of component images.

3. An approach to providing controlled losses

Let us start from considering lossy compression of a single-channel noise-free image in 8-bit representation. After compression, one obtains $\{I_{kij}^c, i = 1, \dots, I, j = 1, \dots, J, k = 1, \dots, K\}$ where quality of this image becomes worse for a larger CR or smaller bpp that takes place for larger QS or SF if a DCT-based coder is applied. Let us see how this happens for JPEG with uniform quantization of DCT coefficients. Suppose that an image to be compressed is divided into N=IJ/4 non-overlapping blocks of the size 8 × 8 pixels. Then, in each block, we have DCT coefficients $\{D(n, k, l), n = 1, \dots, N, k = 1, \dots, 7, l = 1, \dots, 7\}$. After quantization, we have

$\{D_q(n,k,l), n = 1, ..., N, k = 1, ..., 7, l = 1, ..., 7\}$. Then, MSE of losses can be determined as

$$MSE = \frac{1}{N}\sum_{n=1}^{N} MSE_n = \sum_{n=1}^{N}\sum_{k=0}^{7}\sum_{l=0}^{7}\left(\Delta D_q(n,k,l)\right)^2 \quad (4)$$

where

$$D_q(n,k,l) = [D(n,k,l)/QS], k = 0, ..., 7, l = 0, ..., 7,$$

$$\Delta D_q(n,k,l) = QS \times D_q(n,k,l) - D(n,k,l), k = 0, ..., 7, l = 0, ..., 7.$$

and [] denotes rounding-off to the nearest integer, n denotes the block index.

A usual assumption concerning distribution of quantization errors is that it is uniform or close to uniform. Then, MSE is about $QS^2/12$. This is true for quite small QS (see data in **Figure 2**) but, for larger QS, MSE becomes smaller than $QS^2/12$. The main reason is that distributions of alternating current (AC) DCT coefficients differ a lot depending upon an image. **Figure 4** presents these distributions using the same scale for the three considered images (**Figure 1**). Obviously, these distributions differ from Gaussian and from Laplacian (assumed in the paper [33]) as well. For the simple structure image, the distribution is quite narrow and it has heavy tails. If noise is present, the distribution "widens" and becomes closer to Gaussian.

It is seen from analysis of distribution in **Figure 4a** that if QS is about 10, most of AC DCT coefficients become zeros after quantization. Thus, we have decided to analyze quantization errors more in detail. Histograms of these errors for four cases are given in **Figure 5**. The histogram in **Figure 5a** shows that error distribution is close to uniform for the noise-free image Airfield that has wide distribution of AC DCT coefficients (**Figure 4c**). The distribution is also practically uniform for noisy image Frisco (noise standard deviation equals to 5, **Figure 5d**). Then, MSE of introduced losses is really close to $QS^2/12$ (see data in **Figure 2**). In other cases (**Figure 5b** and **c**), the distributions sufficiently differ from uniform. This happens for noise-free image Frisco. Thus, introduced losses MSE is less than $QS^2/12$.

Hence, $MSE \approx QS^2/12$ can be treated as the upper limit of introduced losses. Note that this is valid not only for JPEG but for the coders AGU and ADCT [38–40]. This means that having a desired (threshold) MSE_{des}, it is possible to easily calculate QS as $\sqrt{12\,MSE_{des}}$. A question is when the approximation $MSE \approx QS^2/12$ is valid? Note that if MSE is smaller than $QS^2/12$, one can benefit from using a larger QS and providing a larger CR. Clearly, that if a desired $PSNR_{des}$ has to be provided, it has to be recalculated to MSE_{des} taking into account dynamic range for a given image as $MSE_{des} = D^2/10^{(PSNR_{des}/10)}$.

Our idea [38–40] is that MSE can be predicted in one of two ways.

The first way is determined as

$$MSE_{pred} = \frac{1}{R}\sum_{r=1}^{R} MSE_r = \frac{1}{64R}\sum_{r=1}^{R}\sum_{k=0}^{7}\sum_{l=0}^{7}\left(\Delta D_q(n,k,l)\right)^2 \quad (5)$$

$$\Delta D_q(r,k,l) = QS \times D_q(r,k,l) - D(r,k,l), k = 0, ..., 7, l = 0, ..., 7, r = 1, ..., R \quad (6)$$

where R is the number of analyzed blocks ($R \ll N$), C is a correcting factor used for a given coder. In other words, we employ statistics of DCT coefficients calculated in a limited number R of analyzed blocks of size 8x8 pixels. According to our studies [38, 40], it is enough to have R about 500 where analyzed blocks are

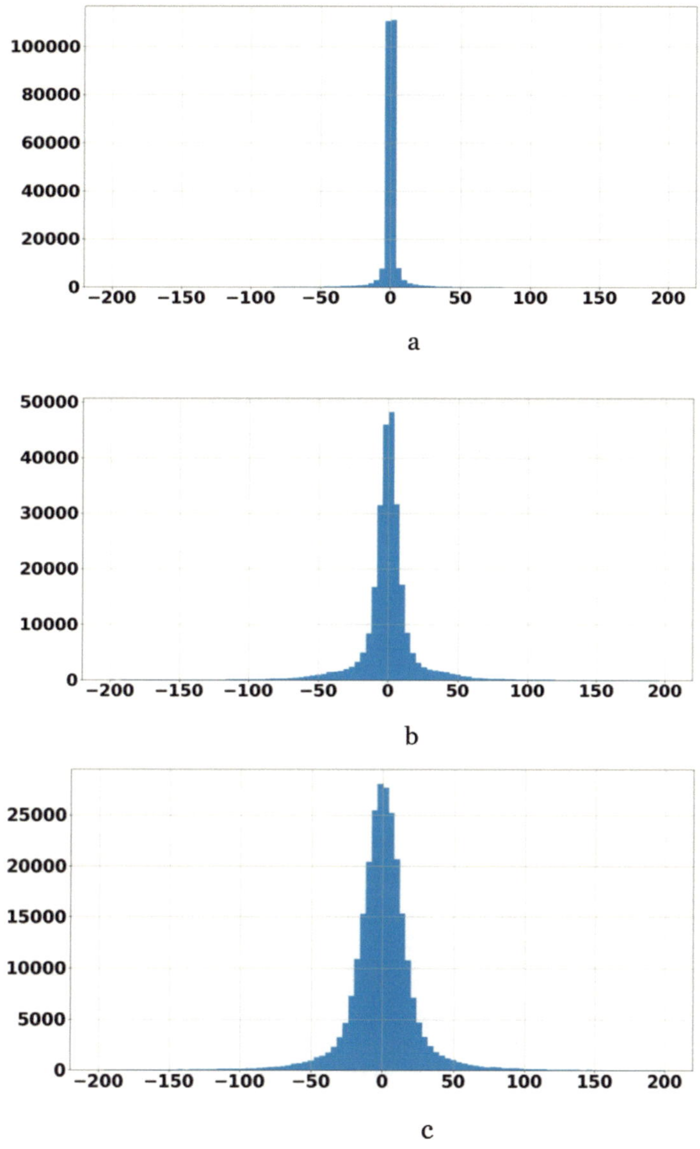

Figure 4.
Distributions of AC DCT coefficients for the noise-free image Frisco (a), noise-free image airfield (b) and noisy image airfield (c), all in the same limits from −200 to 200.

randomly distributed over area of an image to be compressed to have prediction accurate enough. Taking into account that number of 8×8 pixel blocks in compressed images usually exceeds several thousands, prediction occurs to be much faster than even compression by JPEG. Certainly, prediction is much faster than compression by AGU (uses 32×32 blocks, efficient coding and deblocking after decompression) and, especially, ADCT (exploits partition scheme optimization).

Expressions (5 and 6) allow predicting MSE for a given QS. But they do not allow direct setting of QS. One has to apply an iterative procedure that starts from $QS = \sqrt{12\,MSE_{des}}$. If the predicted MSE_{pred} (5) occurs to be considerably (e.g., by 15–20% or more) smaller than MSE_{des}, then a larger QS has to be tried with calculating (6) for all analyzed blocks and (5) again. Since the already calculated DCT coefficients are available, the procedure is quite fast.

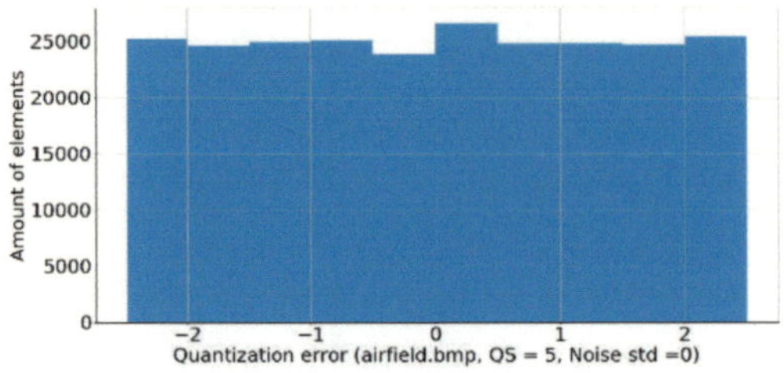

a

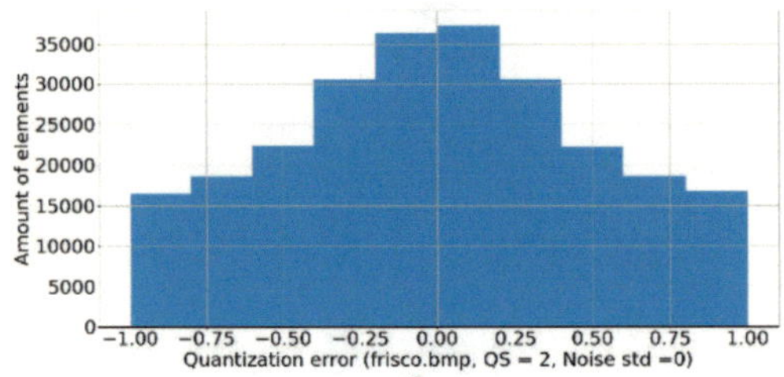

b

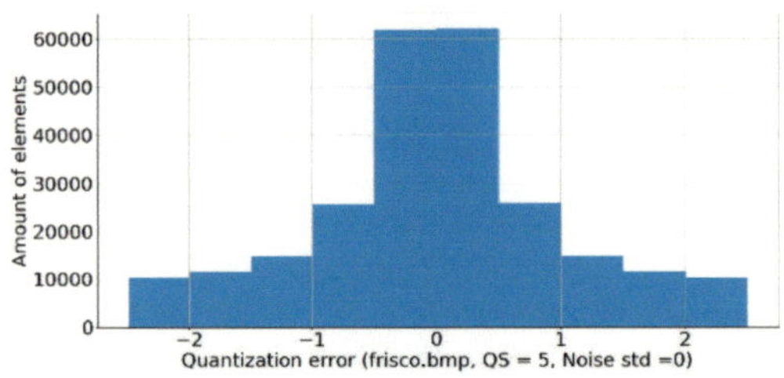

c

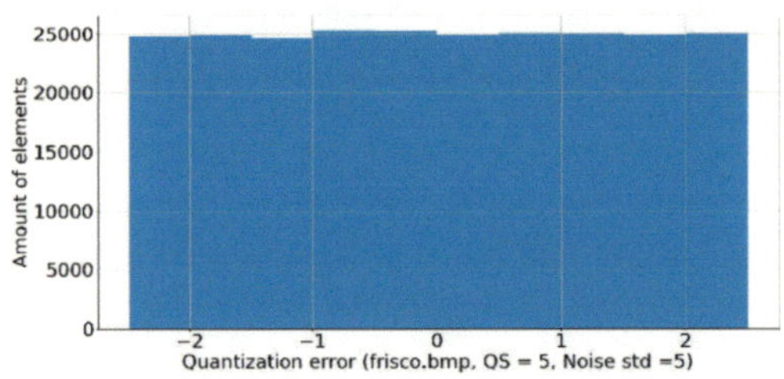

d

Figure 5.
Examples of histograms of quantization error for AC DCT coefficients (see comments under each histogram).

The second way is the following. Suppose that the predicted MSE can be presented as

$$MSE_{pred} = \left(QS^2/12\right) f_0(X) \tag{7}$$

where $f_0(X)$ is a function of one or two parameters X that can be easily and quickly calculated for DCT coefficients determined in analyzed blocks. Then one has to find such parameter(s) and the function. To solve this task, we have exploited our earlier experience in predicting filtering efficiency [49] and compression ratio [18] by simple analysis of DCT statistics in 8x8 pixel blocks and regression analysis [50, 51].

The prediction strategy is the following. We suppose that there is an input parameter (or a few parameters) that can characterize a compressed image. It is also assumed that output (predicted) parameter (MSE, PSNR, CR, or another metric) is strictly connected with this (these) input parameter(s). This connection (prediction approximation) is available to the moment to carry out prediction, i.e., in our case, the function $f_0(X)$ has been obtained in advance (in off-line mode). Then, one has to calculate input parameter(s) for a given QS and insert it (them) into $f_0(X)$.

It has been shown in [52] that a good parameter integrally characterizing an image (its complexity) is probability P_0 that AC DCT coefficients after quantization become equal to zero (this parameter can be also treated as probability that AC DCT coefficient absolute values are smaller than QS/2). It is obvious that P_0 can be very easily calculated. Keeping these properties of P_0 in mind, we have obtained scatter plots of $12MSE/QS^2$ to estimate $f_0(P_0)$. A wide set of test noise-free images has been used that included standard optical images, test RS images and test medical image (this was done to understand does the image nature (origin) influence performance of lossy compression; in fact, very similar results have been obtained for test images of different origin; the main factor is image complexity). Each point of the scatter plot corresponds to one test image compressed with some QS where vertical coordinate is P_0 determined for this case).

Figure 6 presents scatter plots obtained for AGU and ADCT coders with examples of fitted curves. The main and very important observation is that the scatter plots behave in a compact manner, i.e. points that have approximately the same arguments have close values of $12MSE/QS^2$. Another observation is that the scatter plots for two considered coders behave in a very similar manner, i.e. there is a tendency to monotonous decreasing of $12MSE/QS^2$ if P_0 increases. Finally, the scatter plots confirm that, in many practical situations, $MSE \approx QS^2/12$. At least, this is true for $P_0 < 0.6$.

It is worth recalling here that $P_0 < 0.6$ corresponds to rather small QS. To prove this, **Figure 7** presents the scatter plot from [48] and the fitted curve. As it is seen, for $P_0 < 0.6$, CR does not exceed 5. If $P_0 \geq 0.6$, there is the tendency of reduction of $f_0(P_0)$. The scatter plot points are placed not so compactly here. Thus, prediction using only $f_0(P_0)$ becomes less accurate. Nevertheless, the following prediction procedure can be proposed:

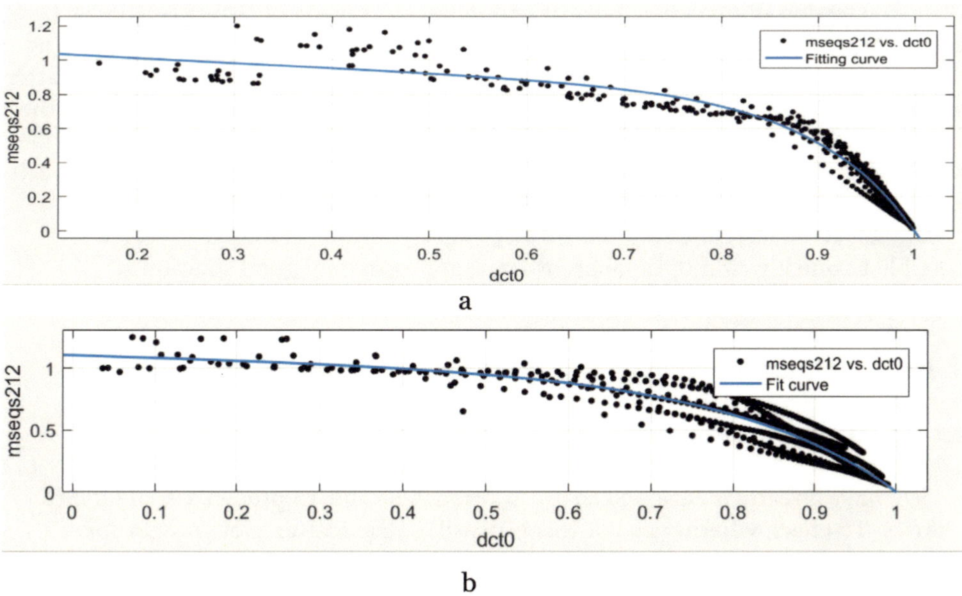

Figure 6.
Scatter plots for AGU (a) and ADCT (b) coders.

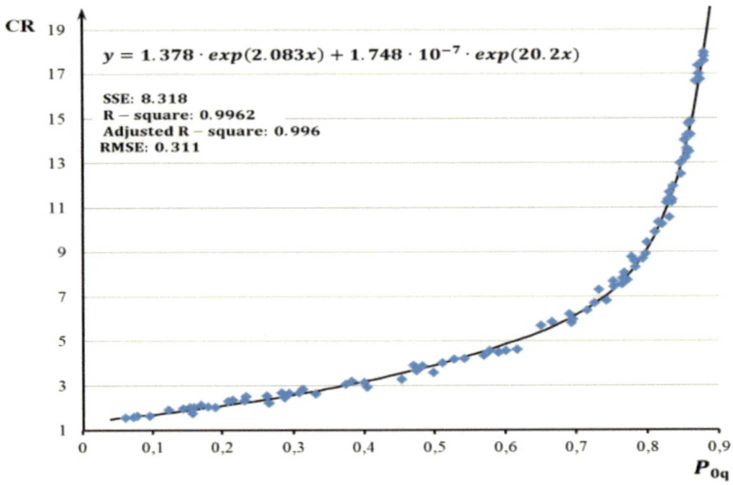

Figure 7.
The scatter plot of CR on Po and the fitted curve for the coder AGU.

1. Determine $QS = \sqrt{12\ MSE_{des}}$, obtain AC DCT coefficients for analyzed blocks and calculate P_0 for this QS.

2. If $P_0 < 0.6$, use $QS = \sqrt{12\ MSE_{des}}$ and stop the procedure.

3. Otherwise, increase QS by about 5%, calculate P_0 and compare $(QS^2/12)f_0(P_0)$ to MSE_{des}; if $(QS^2/12)f_0(P_0) \approx MSE_{des}$ then stop; otherwise continue till satisfying this condition.

As it is seen, all the operations are very easy and fast since they are performed for a limited number of AC DCT coefficients. Moreover, using the same parameter, it is possible to predict both MSE and CR. Then, it is easy to find a proper compromise depending upon priority of requirements and imposed restrictions.

One question is what curves to fit and what are criteria of fitting quality to be used. There are different approaches but we employed goodness-of-the-fit R^2 and RMSE [50] as two main criteria (the former one has to be maximized and the latter one minimized for a given scatter plot). Without going to details, we can state the following. For each scatter plot, usually there are several functions able to provide approximately the same R^2 and RMSE. Sums of two exponentials (see an example in **Figure 7**), polynomials of low order, Fourier series, power functions are good candidates to be tested. Using the corresponding tools of Matlab or Excel, it is possible to quickly find optimal or, at least, appropriately good solution.

4. Peculiarities of compression

4.1 Visual quality metrics

We have already mentioned that it is often desirable to predict visual quality metrics. To check whether or not this is possible, the scatter plot was got for $MSE_{HVS-M}/(QS^2/12)$ vs. P_0 (**Figure 8**). As it is seen, this ratio is about 0.05 for small P_0 (this happens for small QS and/or complex structure images), i.e. PSNR-HVS-M is by about 13 dB larger than PSNR. This means that introduced losses are masked by image content well and, most probably, they cannot be noticed visually.

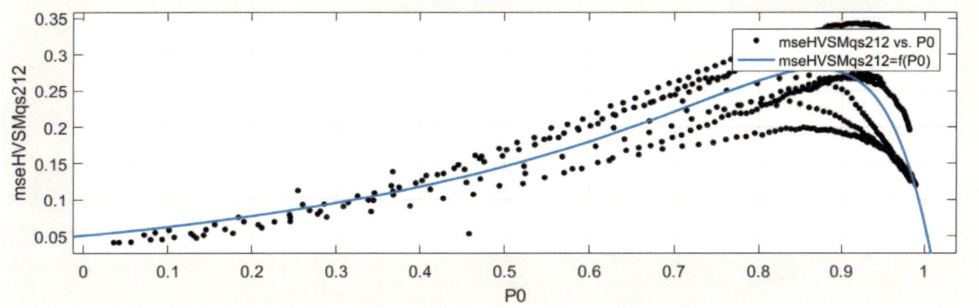

Figure 8.
The scatter plot $MSE_{HVS-M}/(QS^2/12)$ *vs.* P_0 *and the fitted curve, AGU coder.*

The difference in PSNR-HVS-M and PSNR decreases to 5–7 dB for $P_0 > 0.5$, i.e. typical conditions of lossy compression. The scatter plot and the fitted curve show that MSE_{HVS-M} can be predicted well for a given QS. In other words, visual metrics can be predicted too using the proposed approach. Again, the sum of two exponentials (just this case is presented in **Figure 8**) can serve well as approximation curve with quite small number of varied parameters.

4.2 Experimental data for component-wise compression

Let us present the results of applying the proposed approach to real-life hyperspectral data. Images of Hyperion sensor dataset EO1H1800252002116110KZ have been compressed. Hyperion sensor produces data of bad quality (very noisy) in sub-bands with indices k = 1,...,12 and k = 58,...,76. The images in these sub-bands are often discarded in analysis, so we have not compressed them.

Then, two approaches to compression have been compared. Both presume component-wise compression. The first one has been proposed earlier [11]. Images are compressed after applying variance stabilizing transform that takes into account signal-dependent noise properties and converts this noise to additive with variance approximately equal to unity. Then, the recommended QS = 3.5 (this notation is used in figures below). Inverse transform is applied component-wise after decompression. For the proposed method, the component-wise images have been transformed to the interval from 0 to 255. Then, for each of them, AGU coder has been applied with QS = 17 that approximately corresponds to $PSNR_{des} = 34.5$ dB ($MSE_{des} \approx 24 \approx 17 \times 17/12$). The notation QS = 17 is used for the corresponding data.

The obtained PSNR values calculated between compressed and original component images are presented in **Figure 9**. As it is seen, PSNR for the method [11] in most sub-bands occurs to be considerably larger than $PSNR_{des}$ set by us. Only in some sub-bands (indices 165–185) where input PSNR is quite small the determined PSNR values are about 40 dB (i.e., the introduced losses are invisible in decompressed images). For the proposed approach, PSNR for the introduced losses is considerably smaller but, for all sub-band images, PSNR anyway exceeds 35 dB. As it follows from analysis of data in **Figure 10**, CR for all sub-bands exceeds 5 (a more detailed study shows that $P_0 > 0.6$ in all cases). Thus, MSE is smaller than $QS^2/12$ (see data in **Figure 6**) and the provided PSNR is larger than expected.

The main observation for data in **Figure 10** is that CR for the proposed method is by several times larger than for the prototype method for almost all sub-bands except the bands with small input PSNR. Thus, we have gained essential benefit in CR sense while introduced distortions remained invisible.

Figure 9.
PSNR for component-wise compression by the method ([11], QS = 3.5), the proposed component-wise approach (QS = 17), and the proposed 3D compression method (QS = 17, bl = 4).

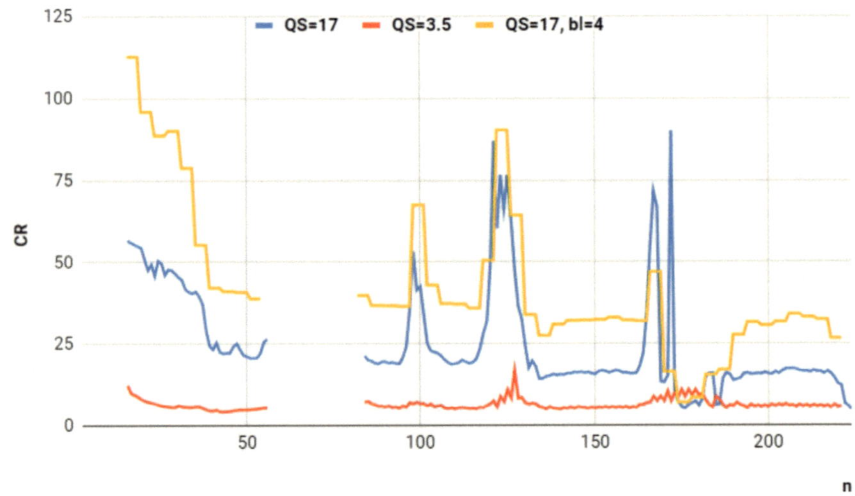

Figure 10.
CR for component-wise compression by the method ([11], QS = 3.5), the proposed approach (QS = 17), and the proposed 3D compression method (QS = 17, bl = 4).

We do not present examples of original and compressed component images because visually they are identical. Note that setting a larger $PSNR_{des}$ leads to larger PSNR of introduced losses and smaller CR for each component image, respectively. By setting a larger $PSNR_{des}$ one can ensure that classification accuracy does not make worse.

4.3 3D compression

Consider now possibilities of 3D compression in groups. There are many different options [11]. We have analyzed one of the simplest ones where component images have been transformed to the 8-bit representation limits, then combined in 4-band groups, and then compressed by 3D version of AGU coder. After decompression the images have to be "stretched" to original limits.

As previously in Section 4.2, we have employed QS = 17. For convenience of comparison, the obtained data are also presented in **Figures 9** and **10**, for 3D compression they are denoted as QS = 17, bl = 4. CR values for the 3D case are shown the same for all components of the same group. As it is seen, CR values for 3D compression are about two times larger than for the proposed component-wise compression. This is an obvious advantage of 3D compression. Meanwhile, there are also very interesting observations stemming from analysis of data for PSNR (**Figure 9**). As it is seen, there are many sub-bands for which PSNR for 3D compression is considerably larger (and the introduced losses are sufficiently smaller) than for component-wise compression. PSNR values are almost the same if sub-bands with small input PSNR are compressed. This is one more positive feature of 3D compression that should be studied more in detail in the future.

5. Conclusions

We have considered the task of lossy compression of RS images with controllable quality characterized by traditional metrics. It is shown that MSE and PSNR can be predicted for DCT-based coders and, due to this, it is possible to provide a desired MSE or PSNR without compression/decompression iterations quite quickly and accurately. Being applied to compress RS images without visible distortions, this approach allows providing CR considerably larger than for approach based on taking noise properties into account.

Moreover, it is demonstrated that prediction of some visual quality metrics is also possible. It is also shown that 3D compression of images collected into groups provides considerably better results. However, additional studies are needed to predict distortion parameters in this case. Examples for real-life data as hyperspectral image are presented.

This research has been partly supported by the Project M/29–2018 of Ukrainian-French program “Dnipro” and STCU Project No. 6386.

Author details

Vladimir Lukin[1*], Alexander Zemliachenko[1], Sergey Krivenko[1], Benoit Vozel[2] and Kacem Chehdi[2]

1 National Aerospace University, Kharkiv, Ukraine

2 University of Rennes 1, Lannion, France

*Address all correspondence to: lukin@ai.kharkov.com

References

[1] Blanes I, Magli E, Serra-Sagrista J. A tutorial on image compression for optical space imaging systems. IEEE Geoscience and Remote Sensing Magazine. 2014:8-26

[2] Schowengerdt R. Remote Sensing: Models and Methods for Image Processing. 3rd ed. Academic Press; 2006. 560 p

[3] Christophe E. Hyperspectral data compression tradeoff in optical remote sensing. In: Prasad S, Bruce LM, Chanussot J, editors. Advances in Signal Processing and Exploitation Techniques. 8th ed. Springer; 2011. pp. 9-29

[4] Yu G, Vladimirova T, Sweeting MN. Image compression systems on board satellites. In: Acta Astronautica. 2009. pp. 988-1005

[5] Magli E, Olmo G, Quacchio E. Optimized onboard lossless and near-lossless compression of hyperspectral data using CALIC. IEEE Geoscience and Remote Sensing Letters. 2004:21-25

[6] Aiazzi B, Alparone L, Baronti S, Lastri C, Selva M. Spectral distortion in Lossy compression of hyperspectral data. Journal of Electrical and Computer Engineering. 2012;**2012**:850637. DOI: 10.1155/2012/850637

[7] Abramov S, Uss M, Abramova V, Lukin V, Vozel B, Chehdi K. On noise properties in hyperspectral images. In: Proceedings of IGARSS; July 2015; Milan, Italy. 2015. pp. 3501-3504

[8] Meola J, Eismann MT, Moses RL, Ash JN. Modeling and estimation of signal-dependent noise in hyperspectral imagery. Applied Optics. 2011: 3829-3846

[9] Uss ML, Vozel B, Lukin V, Chehdi K. Image informative maps for component-wise estimating parameters of signal-dependent noise. Journal of Electronic Imaging. 2013;**22**(1). DOI: 10.1117/1.JEI.22.1.013019

[10] Uss M, Vozel B, Lukin V, Chehdi K. Maximum likelihood estimation of spatially correlated signal-dependent noise in hyperspectral images. Optical Engineering. 2012;**51**(11). DOI: 10.1117/1.OE.51.11.111712

[11] Zemliachenko AN, Kozhemiakin RA, Uss ML, Abramov SK, Ponomarenko NN, Lukin VV, et al. Lossy compression of hyperspectral images based on noise parameters estimation and variance stabilizing transform. Journal of Applied Remote Sensing. 2014;**8**(1):25. DOI: 10.1117/1.JRS.8.083571

[12] Lukin V, Abramov S, Ponomarenko N, Krivenko S, Uss M, Vozel B, et al. Approaches to automatic data processing in hyperspectral remote sensing. Telecommunications and Radio Engineering. 2014;**73**(13):1125-1139

[13] Lukin V, Abramov S, Kozhemiakin R, Vozel B, Djurovic B, Djurovic I. Optimal operation point in 3D DCT-based lossy compression of color and multichannel remote sensing images. Telecommunications and Radio Engineering. 2015;**20**:1803-1821

[14] Zhong P, Wang R. Multiple-spectral-band CRFs for denoising junk bands of hyperspectral imagery. IEEE Transactions on Geoscience and Remote Sensing. 2013:2269-2275

[15] Lukin V, Bataeva E. Challenges in pre-processing multichannel remote sensing terrain images. In: Djurovic I, editor. Importance of GEO Initiatives and Montenegrin Capacities in this Area. The Section for Natural Sciences Book No. 16 Ed. The Montenegrin

Academy of Sciences and Arts Book. No 119. 2012. pp. 63-76

[16] Popov MA, Stankevich SA, Lischenko LP, Lukin VV, Ponomarenko NN. Processing of hyperspectral imagery for contamination detection in urban areas. In: Proceedings of NATO Workshop on Environmental Security and Ecoterrorism; NATO Science for Peace and Security Series C; Springer Science+Business Media B.V. 2011. pp. 147-156

[17] Christophe E, L'eger D, Mailhes C. Quality criteria benchmark for hyperspectral imagery. IEEE Transactions on Geoscience and Remote Sensing. 2005:2103-2114

[18] Zemliachenko A, Abramov S, Lukin V, Vozel B, Chehdi K. Compression ratio prediction in lossy compression of noisy images. In: Proceedings of IGARSS; July 2015; Milan, Italy. 2015. pp. 3497-3500

[19] Christophe E, Mailhes C, Duhamel P. Hyperspectral image compression: Adapting SPIHT and EZW to anisotropic 3-D wavelet coding. IEEE Transactions on Image Processing. 2008;**17**(12):2334-2346

[20] Khelifi F, Bouridane A, Kurugollu F. Joined spectral trees for scalable SPIHT-based multispectral image compression. IEEE Transactions on Multimedia. 2008; **10**(3):316-329

[21] Valsesia D, Magli E. A novel rate control algorithm for onboard predictive coding of multispectral and hyperspectral images. IEEE Transactions on Geoscience and Remote Sensing. 2014;**52**(10):6341-6355

[22] Thayammal S, Silvathy D. Multispectral band image compression using adaptive wavelet transform-tetrolet transform. In: Proceedings of 2014 International Conference on Electronics and Communication Systems; February 2014; Coimbatore, India. 2014. pp. 1-5. DOI: 10.1109/ECS.2014.6892610

[23] Shoba LL, Mohan V, Venkataramani Y. Landsat image compression using lifting scheme. In: Proceedings of International Conference on Communication and Signal Processing; April 2014; India. 2014. pp. 1963-1968

[24] Wang L, Jiao L, Bai J, Wu J. Hyperspectral image compression based on 3D reversible integer lapped transform. Electronic Letters. 2010; **46**(24):1601-1602. DOI: 10.1049/el.2010.1788

[25] Ponomarenko N, Zriakhov M, Lukin V, Kaarna A. Improved grouping and noise cancellation for automatic lossy compression of AVIRIS images. In: Proceedings of ACIVS; Australia; LNCS-6475, Part II. Heidelberg: Springer; 2010. pp. 261-271

[26] Shinoda K, Murakami Y, Yamaguchi M, Ohyama N. Multispectral image compression for spectral and color reproduction based on lossy to lossless coding. In: Proceedings of the SPIE; Image Processing: Algorithms and Systems VIII; February 2010; SPIE 75320H. 2010. DOI: 10.1117/12.838843

[27] Vozel B, Chehdi K, Klaine L, Lukin VV, Abramov SK. Noise identification and estimation of its statistical parameters by using unsupervized variational classification. In: Proceedings of ICASSP; Toulouse, France; vol. II. 2006. pp. 841-844

[28] Bekhtin Yu S. Adaptive wavelet codec for noisy image compression. In: Proceedings of the 9th East-West Design and Test Symp.; Sept., 2011; Sevastopol, Ukraine. 2011. pp. 184-188

[29] Al-Chaykh OK, Mersereau RM. Lossy compression of noisy images.

IEEE Transactions on Image Processing. 1998;7(12):1641-1652

[30] Kozhemiakin R, Abramov S, Lukin V, Djurović I, Vozel B. Peculiarities of 3D compression of noisy multichannel images. In: Proceedings of MECO; June 2015; Budva, Montenegro. 2015. pp. 331-334

[31] Kozhemiakin RA, Zemliachenko AN, Lukin VV, Abramov SK, Vozel B. An approach to prediction and providing of compression ratio for DCT-based coder applied to remote sensing images. Ukrainian Journal of Earth Remote Sensing. 2016;**8**:22-29

[32] Jiang H, Yang K, Liu T, Zhang Y. Quality prediction of DWT-based compression for remote sensing image using multiscale and multilevel differences assessment metric. Mathematical Problems in Engineering. 2014;**2014**:15 Article ID 593213

[33] Minguillon J, Pujol J. JPEG standard uniform quantization error modeling with applications to sequential and progressive operation modes. Electronic Imaging. 2001;**10**(2):475-485

[34] Ponomarenko NN, Lukin VV, Egiazarian K, Astola JDCT. Based high quality image compression. In: Proceedings of 14th Scandinavian Conference on Image Analysis; Joensuu, Finland. 2005. pp. 1177-1185

[35] Ponomarenko N, Lukin V, Egiazarian K, Astola JADCT. A new high quality DCT based coder for lossy image compression. In: CD ROM Proceedings of LNLA; August 2008; Switzerland. 2008. p. 6

[36] Zemliachenko A, Ponomarenko N, Lukin V, Egiazarian K, Astola J. Still image/video frame lossy compression providing a desired visual quality. Multidimensional Systems and Signal Processing. June 2015:22. DOI: 10.1007/s11045-015-0333-8

[37] Kozhemiakin R, Lukin V, Vozel B. Image quality prediction for DCT-based compression. In: Proceedings of CADSM 2017; Ukraine. February 2017. pp. 225-228. DOI: 10.1109/CADSM.2017.7916121

[38] Vozel B, Kozhemiakin R, Abramov S, Lukin V, Chehdi K. Output MSE and PSNR prediction in DCT-based lossy compression of remote sensing images. In: Proceedings of the SPIE. 10427, Image and Signal Processing for Remote Sensing XXIII; Warsaw, Poland. September 2017. p. 11

[39] Krivenko S, Zriakhov M, Lukin V, Vozel B. MSE prediction in DCT-based lossy compression of noise-free and noisy remote sensing images. In: Proceedings of TCSET; Lviv-Slavske, Ukraine. February 2018. p. 6. DOI: 10.1109/TCSET.2018.8336338

[40] Krivenko S, Lukin V, Vozel B. MSE and PSNR prediction for ADCT coder applied to lossy image compression. In: Proceedings of The 9th IEEE International Conference on Dependable Systems, Services and Technologies DESSERT'2018; Kiev, Ukraine. May 2018. p. 6. DOI: 10.1109/DESSERT.2018.8409205

[41] Taubman D, Marcellin M. JPEG2000 Image Compression Fundamentals, Standards and Practice. 1st ed. Springer; 2002. DOI: 10.1007/978-1-4615-0799-4

[42] Zemliachenko AN, Abramov SK, Lukin VV, Vozel B, Chehdi K. Lossy compression of noisy remote sensing images with prediction of optimal operation point existence and parameters. Journal of Applied Remote Sensing. 2015;**9**(1):095066. DOI: 10.1117/1.JRS.9.095066

[43] Rissanen J. Modeling by shortest data description. Automatica. 1978; **14**(5):465-471. DOI: 10.1016/0005-1098(78)90005-5

[44] Rubel O, Zemliachenko A, Abramov S, Krivenko S, Kozhemiakin R, Lukin V, et al. Processing of multichannel remote-sensing images with prediction of performance parameters, chapter 13. In: Environmental Applications of Remote Sensing. Intech; June 2016. pp. 373-416

[45] Green RO, Eastwood ML, Sarture CM, Chrien TG, Aronsson M, Chippendale BJ, et al. Imaging spectroscopy and the airborne visible/infrared imaging spectrometer (AVIRIS). Remote Sensing of Environment. 1998;**65**:227-248

[46] Pearlman JS, Barry PS, Segal CC, Shepanski J, Beiso D, Carman SL. Hyperion, a space-based imaging spectrometer. IEEE Transactions on Geoscience and Remote Sensing. 2003: 1160-1173. DOI: 10.1109/TGRS.2003.815018

[47] Ponomarenko N, Silvestri F, Egiazarian K, Carli M, Astola J, Lukin V. On between-coefficient contrast masking of DCT basis functions. In: CD-ROM Proceedings of VPQM; USA. 2007. 4 p

[48] Lukin V, Ponomarenko N, Egiazarian K, Astola J. Analysis of HVS-metrics' properties using color image database TID2013. In: Proceedings of ACIVS; Italy. 2015. pp. 613-624

[49] Abramov S, Krivenko S, Roenko A, Lukin V, Djurovic I, Chobanu M. Prediction of filtering efficiency for DCT-based image denoising. In: Proceedings of MECO; June 2013; Budva, Montenegro. 2013. pp. 97-100

[50] Cameron C, Windmeijer A, Frank AG, Gramajo H, Cane DE, Khosla C. An R-squared measure of goodness of fit for some common nonlinear regression models. Journal of Econometrics. 1997; **77**:329-342

[51] Rubel O, Abramov S, Lukin V, Egiazarian K, Vozel B, Pogrebnyak A. Is texture denoising efficiency predictable. International Journal on Pattern Recognition and Artificial Intelligence. 2018;**32**. DOI: 10.1142/S0218001418600054

[52] Zemliachenko A, Abramov S, Lukin V, Vozel B, Chehdi K. Improved compression ratio prediction in DCT-based lossy compression of remote sensing images. In: Proceedings of IGARSS; Beijing, China. 2016. 4 p. DOI: 10.1109/IGARSS.2016.7730817

Section 3

Ionospheric Monitoring Systems

Chapter 4

Reverse Satellite Transionospheric Sounding: Advantages and Prospects

Igor Ivanov, Olga Maltseva, Vladimir Sotskii, Alexandr Tertyshnikov and Gennadii Zhbankov

Abstract

This chapter includes four sections. The first introduction section provides a brief review of the existing methods of transionospheric sounding and the results obtained, and the shortcomings of each are noted. The second section describes the proposed principle based on the installation of a receiver on the GLONASS platform. The advantages and technical characteristics of the proposed system are justified. The main area of use is the polar region. The third section presents the most modern modeling methods and models used. To calculate the propagation of radio waves, this is a method of ray tracing taking into account the large- and small-scale inhomogeneities of the ionosphere. To describe the state of the ionosphere, it is proposed to use the IRI2016 model, which includes adaptation to the current diagnostic data provided by ground ionosondes, and the IRI-Plas model, which not only can be adapted to ground ionosonde data, but also to values of the total electronic content, the measurement of which is an additional advantage of the proposed system. The fourth sectionincludes areas of application, the main of which is the monitoring of the polar region, and the least provided with ionospheric information.

Keywords: satellite, satellite information, transionospheric sounding, modeling, radio wave propagation

1. Introduction: a brief review of existing methods of transionospheric sounding

The role of the ionosphere in providing human life is difficult to overestimate. This role was great in the pre-satellite era and multiplied after the launch of an artificial Earth satellite. This role is most fully represented in [1], in which various technological systems are divided into two categories: (1) systems which cannot exist without the ionosphere (VLF-LF communication and navigation, MF communication, HF communication, "short-wave" listening, OTH radar surveillance, HFDF and HF SIGINT), and systems on which the ionosphere makes the big impact (a satellite communication, satellite navigation (i.e., GPS & GLONASS, etc.), space-based radar and imaging, terrestrial radar surveillance and tracking, and others). This shows how it is important to study the ionosphere. Classical methods of studying the ionosphere from the Earth's surface are the pulsed sounding—the sending of radio pulses and the observation of their reflections from various ionospheric

IntechOpen

layers with the measurement of the delay time and the study of the intensity and shape of the reflected signals. Measuring the reflection height of radio pulses at different frequencies, determining the critical frequencies of different regions, that is, frequencies for which the given area of the ionosphere becomes transparent, it is possible to determine the value of the electron density in the layers and the virtual altitudes for the given frequencies, hence to select the optimum frequencies for the specified radio paths. This is the principle of work of the main device namely of ionosonde. Its use for many years in many parts of the globe has made it possible to obtain a huge array of data and develop predictive models of various parameters. The breakthrough occurred with the advent of satellites, in particular, with the installation of ionosondes on satellites [2]. Here, there is a wide variety of methods. Let us list some of them with a brief description. The first is external (topside) sounding (TS, f < foF2, where foF2 is the critical frequency of the basic layer of the ionosphere F2). An example is the on-board ionosonde "IS-338," in future projects, the on-board ionosonde "LAERT" [3]. The second is direct transionospheric sounding (TIS) (f > foF2). In this case, the signal is emitted from the spacecraft, and the reception is on the Earth. The third option is reverse transionospheric sounding (RTS) (f > foF2). In this case, the signals emitted by the transmitter from the Earth are received by the receiver on the satellite. All of these methods were implemented in a variety of experiments on spacecraft series Alouette, Arial, ISIS, Intercosmos-19 and Kosmos-1809, IC Bulgaria 1300, and others. An example of ionograms of TS (the left curve) and RTS (the right curve) of "Kosmos-1809" is shown in **Figure 1**.

A detailed analysis of the current state of the TS method is given in [4]. The RTS method is presented in [5]. In all cases, a number of important tasks were solved. One of them is the mathematical support of methods [6]. The second problem is the synchronization of airborne and ground ionosondes [7]. **Figure 2** shows a system for synchronous ionospheric sounding using all methods of radio pulse sounding. The features of synchronous operation of the terrestrial-satellite system are presented in the examples of synchronization schemes for previously launched Russian ionosondes.

As to scientific results, it is difficult to select something from hundreds of, if not from thousands of, the publications based on processing of millions of ionograms. As examples, it is possible to note the first publications [8] and the detailed review of results from Alouette1, Explorer 20, Alouette2, Explorer 31 [9]. Data of topside sounding have played a big role in improvement of various models of the ionosphere with the use of data of Intercosmos-19 and Cosmos 1809 [10], Alouette1-2, ISIS-1-2 [11] and later satellites [12]. In the paper [13], application of ionospheric

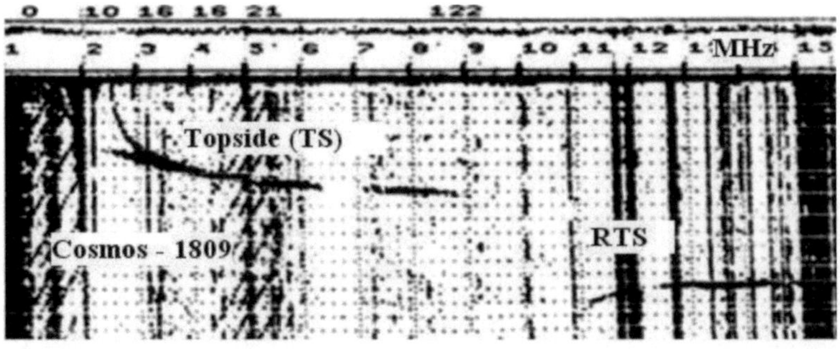

Figure 1.
Example of ionograms of TS and RTS sounding.

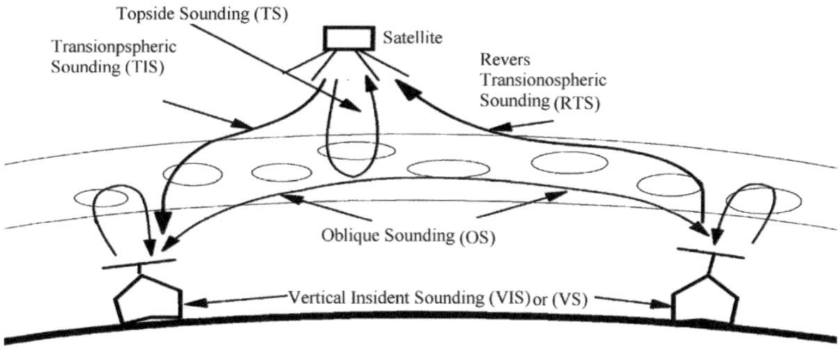

Figure 2.
Scheme of system sounding of the ionosphere.

topside-sounding results to magnetospheric physics and astrophysics is presented. From the latest publications, it is possible to specify the study of behavior of high-latitude N(h)-profiles during strong magnetic storms with use of huge database of digital topside ionograms [14]. It is underlined in [4] that the interest in the modification and the use of TIS has increased recently in connection with the need to monitor the polar region. On the one hand, this is due to the fact that high-latitude regions were always problematic zones because of the small number of ionosondes and the enormous spatial and temporal variability of the ionosphere due to the magnetospheric influence. On the other hand, interest increased due to increased scientific and economic activities in these regions.

In this chapter, the main emphasis is also on the study of the polar region, that is why we consider the existing proposals. In [15], the authors, noting such shortcomings of the TIS method, using low-altitude satellites, associated with the high speed of the satellite and its projection to the Earth's surface, as impossibility of exact separation of the spatiotemporal characteristics of the variability in a diagnosed region and large time delay between the measurements of the ionospheric parameters and the processing and analysis of the measurement data, have proposed the use of satellites in the geostationary orbit. Since it is proposed to install a transmitter on a satellite, the main attention in [15] is given to the energy problem. A calculation of ray trajectories is carried out in a two-dimensional plane of propagation from a satellite to a receiver point formed by combined vertical profiles of the electron density. Up to a height of 2000 km, Ne(h) is taken from the IRI model and then is "sewed" to the NeQuick model [16] in such a way that the electron density becomes zero at a height of 36,000 km. Calculations were made for the transmitting antennas as weakly directed vertical or horizontal dipoles. Of the three types of signals: simple smooth pulses, linearly frequency modulated (LFM) and phase and code manipulated (PCM) used for sounding the ionosphere, estimates are made for the first and second types for signal-to-noise ratio (S/N) of 20 dB. For the first type, the field strength at the receiver was ~6 μB/m for radiated power of 1 kW. The use of LFM signals of ~100 μs duration is intended to reduce the radiated power due to the complexity and broadband of the signals. At S/N = 20 dB, the radiated power can be 100 W. An example of the model transionogram and the received field strength is shown in **Figure 3**.

The paper [17] is a definite addition to [15], of which deserves an attention remark about another disadvantage of this method TIS and RTS, namely, that these techniques only determine two parameters: the critical frequency (foF2) and the height of the maximum of the F2 layer (hmF2) in the sub-satellite point, without defining N(h)-profile, and that to date conducted only one successful

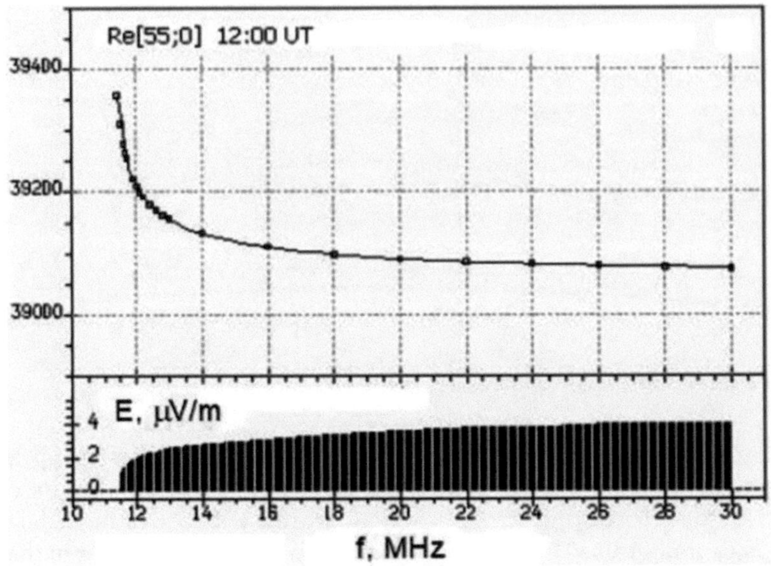

Figure 3.
Illustration of the sounding method from a geostationary satellite at the point of reception 55° N during the daytime (UT = 12).

experiment about reconstruction of a bottom Ne(h)-profile of the F2 layer using them [18]. It is possible that namely for this reason, the methods of TIS and RTS have not been developed in the world. Explanation was made for differences of the proposed methods of TS, TIS and RTS: (a) a source of information is not a traditional ionogram and ionogram, retaining only radio sounding signals; (b) ionosonde becomes a measuring tool, and the system of a single-board broadband transmitter module (BTM) and a network of ground receiving k modules (GRM) with the tuning frequency tunable synchronously with the on-board BTM. Therefore, the method is called as multifrequency ionospheric radioscopy method. The difference between [17] and [15] is information on experimental data confirming the possibility of implementing the proposed method, but these data are indirect. The approach described in [19] is based on real experiments in the Arctic using external sounding [20]. These experiments revealed a shortcoming of use of circular low orbit, namely the inability to determine the dynamics of the ionospheric irregularities; therefore, the use of space satellites with highly elliptical orbits with an apogee over the North Pole was proposed. However, it was immediately noted the complexity of using such orbits, consisting in the difficulty of choosing the time of ionosonde location over the investigated area. To overcome this difficulty, it is proposed to use solar-synchronous orbits. It is shown that there is a number of such orbits for monitoring Arctic of Russia: the first orbit has an apogee of 40,000 km (the North Pole) and a perigee of 500 km (above the South Pole) during one revolution of 12 h, a second orbit has an apogee of 20,343 km, perigee 485 km, the time of one revolution is 6 h, the third orbit has apogee 11,829 km, perigee 500 km. The choice of the orbit is mainly influenced by the significant energy loss of the probing signal for propagation in free space. For example, the difference in signal power loss from reflection from the Earth between a high-orbit and low-orbit ionosondes can be about 100 dB, and the transmitter power is limited. As a solution, signals with phase-code modulation are selected. Calculation of the sounding parameters has shown that a power of 600 W will not suppress the remaining experiments on the satellite, and at the same time, this power will be enough to assure confident reception of signals reflected from the ionosphere of the Earth on a high-altitude apogee satellite. All these proposals indicate the relevance and prospects of such methods.

2. The proposed principle

The main element of the methods [15, 17, 19] is the installation of an ionosonde on the spacecraft, but as noted in [21], an ionosonde for TS [19] or a powerful on-board transmitter for TIS [22] not yet created. Even more important is the problem of electromagnetic compatibility on board of a satellite. The proposed principle is a kind of RTS and consists of the use of an on-board receiver and a terrestrial transmitter, which makes it possible to increase the energy potential of the sounding channel due to a more powerful ground-based transmitter. The author of [21] showed the possibility of RTS from highly elliptical orbits up to 40,000 km. The details are as follows. It is proposed to use a terrestrial ionosonde of the "Parus" type with a typical rhombic antenna, only an ionosonde receiver with shortened antennas should be installed on board (the upper part of the frequency range is used). For comparison, the calculation is performed for the orbits proposed in [19], and the characteristics of which are given in **Table 1**. The orbital number, apogee and perigee, eccentricity and period of revolution are given.

The possibility of successful sounding of the ionosphere by means of satellites is largely determined by the energy potential of the Pv channel of the satellite-Earth. The parameter Pv is estimated in [19] for TS and in [17] for TIS. To assess Pv, in the case of RTS, ground and airborne antennas are used, the directional patterns and gain factors are shown in **Figures 4** and **5**. The tables, included in the form of frames, give an idea of the quantitative evaluations: Freq—frequency, R—active resistance, jX—reactive resistance, Ga—gain, F/B—ratio of direct radiation and reverse power, and ON—auxiliary symbol.

The characteristics of the assumed on-board receiver and ground transmitter are given in **Table 2**.

The energy potential of Pv with RTS for orbits with a height of 10–40,000 km was calculated according to the known relation from [23]. The results are shown in **Figure 6**. The red line presents the calculation results for the frequency of 5 MHz, green—10 MHz, blue—15 MHz, dark blue—20 MHz.

In the paper [22], the calculation of the power of the communication channel was carried out for TIS in the circumpolar region, and it is shown that "the main source of background noise is the radiation of powerful radio stations on the Earth's surface." That is why, for comparison, **Figure 5**, the dashed curve shows the level of interference from a 21-m broadcast transmitter with a power of 50 kW at the altitudes of the satellite. The peculiarity of the on-board ionosonde "Laert" is a two-channel polarization reception, which allows reducing to a minimum the losses from the polarization mismatch. Losses in antenna feeders

Orbital number	Apogee (km) above the North Pole	Perigee (km)	Eccentricity	Circulation period (h)
1	40,000	500	0.74	12
2	20,343	485	0.591	6
3	11,829	500	0.468	4

Table 1.
Characteristics of the supposed orbits of satellites.

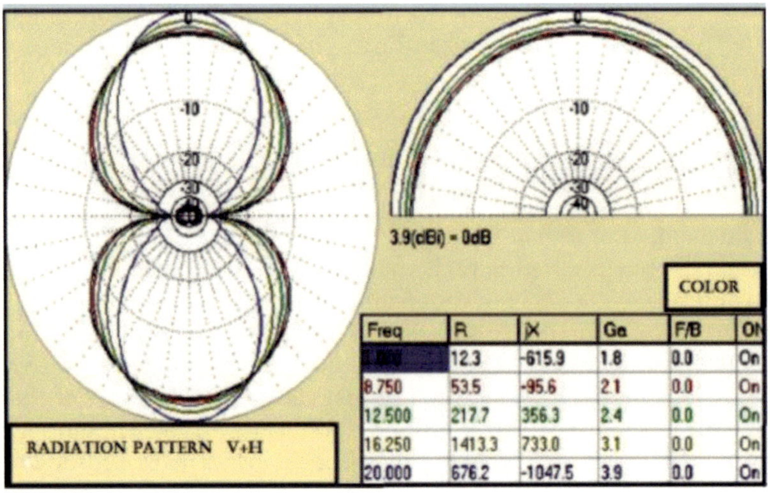

Figure 4.
The calculated radiation patterns and gain factors for on-board orthogonal antennas of 15 m length analogous to the ionosonde antennas of Cosmos-1809.

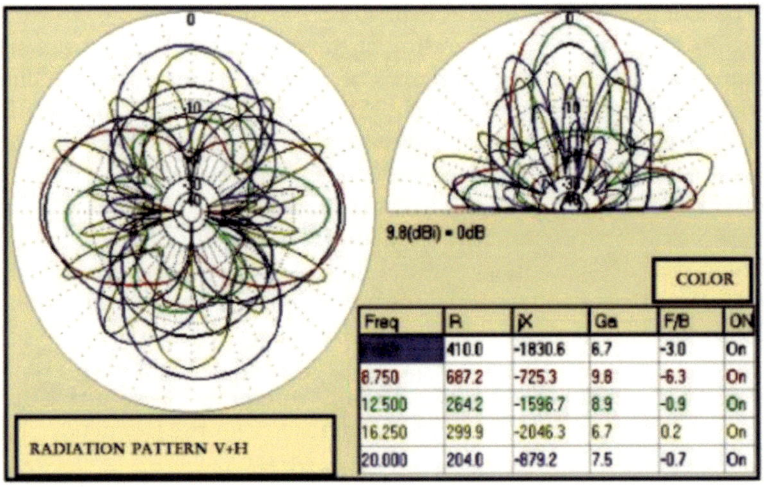

Figure 5.
The directional pattern and the amplification factor of the terrestrial vertical orthorhombic antenna of the VS ionosondes.

and filters are assumed to be 3–6 dB. The obtained values of Pv show the excess of the signal above the interference to 6–10 dB even for distances exceeding 40,000 km. The worst reception conditions are near the cutoff frequency. In the case under consideration, this frequency of 5 MHz corresponds to foF2, and it is possible to use a corresponding estimation of absorption. However, in this case too, at a low signal-to-noise ratio (S/N), the transionograms can be recorded, as was shown in the experiments "Intercosmos-19" and RTS—with "Cosmos-1809." The RTS sounding mode, in which one pulse is emitted at each frequency, ensures maximum diagnostic efficiency, but the "Parus-A" and "Laert" ionosondes are potentially designed to work with complex signals, which can be used to increase the S/N ratio during a significant increase of the absorption in the polar ionosphere during different disturbances. Thus, the RTS mode is energetically favorable for the sounding channel, and it does not require, as with the TIS

Parameter	Receiver of the ionosonde "Laert"	Transmitter ionosonde "Parus-A"	Notes
Range of sounding frequencies (MHz)	0.1–20	0.5–20	Installed programmatically
Pulse power of the transmitter (kW)		Up to 20	Lamp version
Pulse width (μs)	100	20–200	
Pulse repetition frequency (Hz)	60	50–100	Installed programmatically
Bandwidth (kHz)	15		(−3 dB)
Sensitivity (μV)	Not less than 5		With S/N 10 dB
Dynamic range (dB)	Not less than 100		Up to 120
Number of discrete frequencies	400	Not less than 400	Changes programmatically
The law of adjustment in the range	Logarithmic	Linear	Changes programmatically
Instability of reference frequency	Not worse than 10^{-8}	Not worse than 10^{-8}	

Table 2.
Characteristics of prospective receiving and transmitting devices.

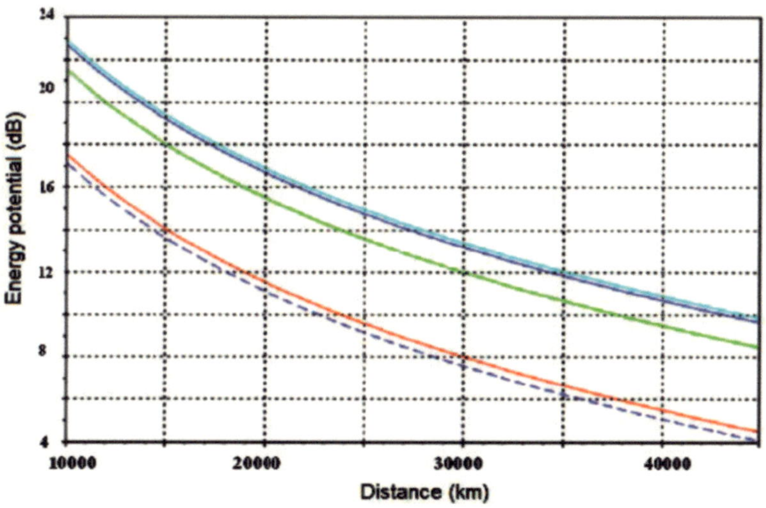

Figure 6.
The energy potential of the Pv communication channel at the RTS for orbits with a height of 10–40,000 km. Red line shows results for frequency 5 MHz, green—for 10 MHz, blue—for 15 MHz, dark blue—for 20 MHz.

in [22], to install on the satellite a powerful transmitter that affects the operation of other on-board systems to ensure the requirements of the electromagnetic compatibility. Obviously, working with complex signals, with the accumulation of signals at each sampling frequency [19], significantly reduces the resolution concerning space. The synchronization problem can be solved with the "binding" of the chronographs of the ground and airborne ionosondes to the exact time of GLONASS or GPS [17, 21], taking into account the ephemerides of the satellite.

These results allow us to propose a somewhat different scheme of the RTS method, namely, the use of the GLONASS platform to install the receiver (an application of this method is given in Section 4).

3. Used modeling methods and models

3.1 Ionospheric models

Among large number of empirical models of the ionosphere, the International Reference Ionosphere (IRI) is the most widely used, tested by huge scientific community and constantly modified. The development of this model was started in the late 1960s and was carried out under the umbrella of Committee on Space Research COSPAR and International Union of Radio Science URSI. Currently, it is an international standard for the determination of ionospheric parameters [24]. This is a statistical average model based on a huge amount of data from both terrestrial and satellite measurements. For the propagation problems, its most important parameters are: the critical frequency foF2 of the F2 layer (or the maximum density NmF2 bound by the linear ratio with the square of the critical frequency), the height hmF2 of the maximum of the F2 layer, the propagation coefficient M3000F2 determining the maximum applicable MUF frequency for the 3000 km path, the altitude profile of the electron density N(h), the total electron content TEC of the ionosphere. The parameters are determined using the coefficients of CCIR and URSI, obtained by the Fourier expansion according to the data of the 1960s and the 1980s. The driving parameters are the solar activity indices. The input parameters are the date, latitude and longitude of the point on the globe. The shape of the N(h)-profile of the lower part of the ionosphere is determined by parameters B0 and B1, for which there are two options: tabular and Gulyaeva data, but they do not have advantages over each other, although the tabular variant is most often used. There are several basic versions of the model, reflecting the most significant stages of its modification: IRI79, IRI90, IRI95, IRI2001, and IRI2007. The latest modifications of the IRI2016 model are presented in [25]. They include two new options for hmF2. The review of the last steps for the transition from climatological character of the model to the description of ionospheric conditions in real time by adaptation to the current diagnostic data is presented. At present, there is a new version of IRI-Plas [26]. The main distinguishing features of this model are: (1) introduction of a new height scale for the upper ionosphere, (2) taking into account the plasma sphere part of the profile and (3) adapting the profile to the experimental value of TEC. The advantages provided by these features are indicated in [27].

3.2 Trajectory calculation method

Like a situation with ionospheric modeling, there is a number of methods of radiowave propagation calculation in the model ionosphere. In this chapter, the most theoretically developed method of trajectory calculations is used. Briefly, it consists of the following. In general, the calculation of trajectories follows the classical ray tracing procedure [28]. A numerical solution of the local dispersion equation is found by transforming it to a system of differential characteristic equations with respect to spatial and ray coordinates in the model ionosphere as the sum of the basic unperturbed part and the additional perturbation: $N_e = N_0 \cdot (1 + \delta_L + \delta_T)$ where N_0 is the unperturbed "base" part described by the international model IRI-2016, with allowance for the possibility of correction in the presence of experimental vertical sounding data, δ_L and δ_T are the perturbations created by traveling ionospheric disturbances (TID).

3.3 Determination of the parameters of inhomogeneities

Using the proposed method, RTS allows to study in detail not only the large-scale structure in the form of layers, but also to determine the parameters of quasi-wave disturbances, including traveling ionospheric disturbances (TIDs) in the outer ionosphere. The technique was proposed and tested according to the data of RTS on the Intercosmos-1809 satellite. Details of the method are presented in [29]. It is the singular spectrum analysis (SSA) method [30], which is modified to select a one-dimensional latitudinal series of observations of the quasi-harmonic TID component at a fixed height. The investigated series are the values of the electron density at a fixed height in the function of the geomagnetic co-latitude. These series are decomposed into a sum of series, each of which corresponds to a trend, a periodic component and, possibly, noise. The basic algorithm of the SSA method includes four steps: embedding, singular decomposition, grouping, and diagonal averaging. Modification of the basic SSA method consists of removing the trend and analyzing the remaining term characterizing the disturbance. The main contribution is given by decomposition [31]. The method was applied to the case of ionospheric sounding from the Kosmos-1809 satellite on 06.03.1987, pass 1079, UT = 8.82–8.92 h along geomagnetic longitude ~128° E. The geomagnetic situation during the experiment was characterized by the planetary index Kp = 2+. The spatial distribution of the normalized values of the electron density in the altitude range of 300–400 km is shown in **Figure 7**.

Figures 8 and **9** show the results of the determination of the electron density perturbation at an altitude of 350 km.

Figure 8 shows that the disturbance is a fragment of a wave train, not a harmonic. This clearly manifests itself when considering the spatial picture of the disturbance depicted in **Figure 9**. In this particular case, we are dealing with a weak perturbation with a wavelength of L ≈ 500 km, damped in the vertical and horizontal directions.

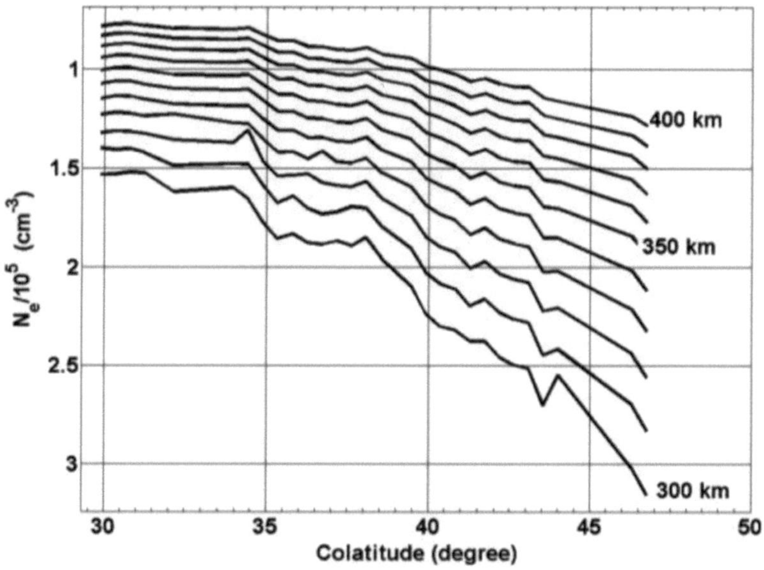

Figure 7.
Spatial distribution of normalized values of the electron density from topside radio sounding of the ionosphere from the satellite Kosmos-1809.

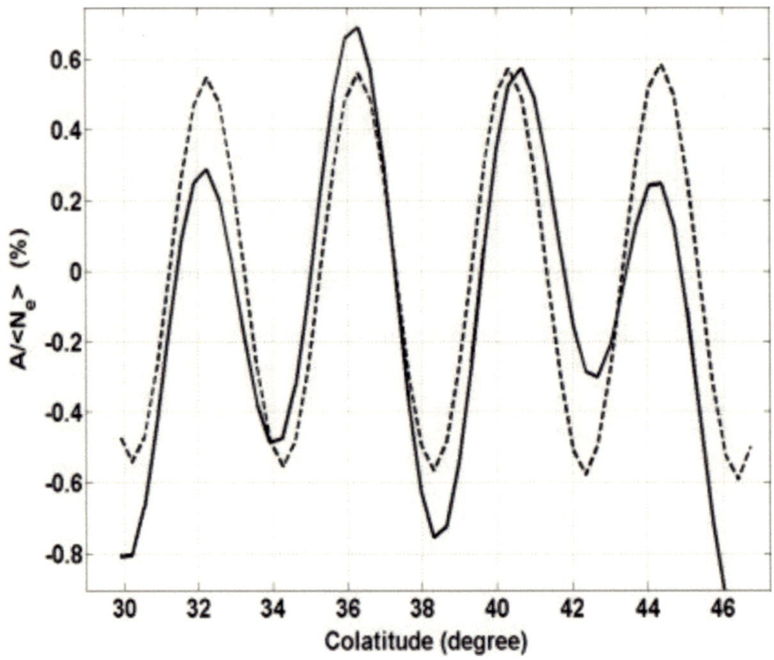

Figure 8.
The results of the analysis of experimental data for the height of 350 km: a quasi-harmonic component of the transformed normalized number (solid line) and its approximation damped sinusoid (dotted line).

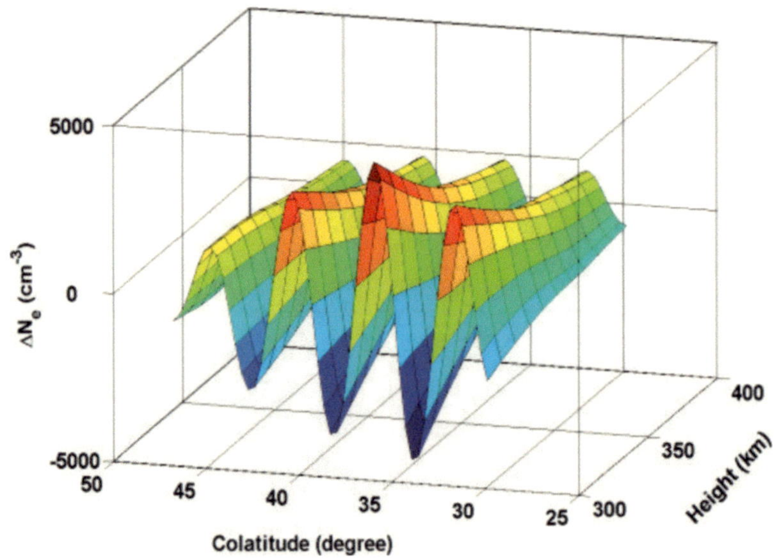

Figure 9.
The spatial pattern of quasi-wave perturbations obtaining in the application of the modified SSA method to the analysis of experimental data of the topside radio sounding ionosphere on board of the satellite "Cosmos-1809" 1987.03.06, at 8.82–8.92 UT.

3.4 Use of TEC

TEC measurements by means of signals from satellites began literally with the launch of the first satellite [32]. Since the mid-1960s, continuous measurements of the rotation of the Faraday polarization plane at one point with the help of geostationary satellites have been carried out. Since the mid-1970s, measurements

DOI: http://dx.doi.org/10.5772/intechopen.80240

have been carried out using differential group delays and Doppler shifts. As noted by many researchers, the TEC parameter, like the critical frequency foF2, has become the main parameter for describing the behavior of the ionosphere. A huge number of GPS receivers around the globe, the availability of data on the Internet has allowed and allow scientists to study the local, regional, global characteristics of the ionosphere independently, quickly and simultaneously. The results are presented both on the basis of the data of local networks of GPS receivers, and data of global maps. In this chapter, the TEC values are calculated from the IONEX file data (site ftp://cddis.gsfc.nasa.gov/pub/gps/products/ionex/) for the global JPL map.

3.5 Examples of using models

Since any empirical model is problematic for the high latitude region, it is necessary to obtain quantitative estimates of the accuracy of the model in the investigated region. In our case, it is the European part of Russia. **Figure 10** gives an example of a comparison of the model and experimental ionospheric parameters for the highest latitude Longyearbuen station (78.2° N, 15.9° E). Its results are compared with the results of the mid-latitude station Juliusruh, which is a reference station. Results are given for mean solar activity from those vertical sounding data that were available for the Longyearbuen station (2011–2014). Absolute |ΔfoF2| (in MHz) and relative deviations σ (in %) are compared for three options: (1) model values and experimental medians ("med" icon), (2) model values and monthly average experimental instantaneous values ("ins"), and (3) values calculated using experimental TEC and equivalent slab thickness of the ionosphere τ ("τ" icon). The latter should be compared with option 2.

The values averaged over the year during the data availability period of Longyearbuen station are given in **Table 3**. Each result column contains |ΔfoF2| and σ. **Figure 9** and **Table 3** show that the average deviations for options 1 and 2 for both stations are not much different. This indicates that the IRI model can provide foF2 with accuracy close to mid-latitude values. Using TEC improves the correspondence between calculated and experimental values in 1.5–2 times; however, here, the results for Longyearbuen station are worse than for Juliusruh. The reason lies in the small number of stations contributing to the construction of global TEC maps, and at large angles of slant TECs associated with the boundary latitude of navigation satellites.

An interesting result was obtained by the author of the model SDMF2 (Satellite and Digisonde Data Model of the F2 layer) for quiet geomagnetic conditions [33] when compared with the IRI model [34]. A comparison was made using data from eight ionosondes in various regions of the globe to evaluate the effects of different solar-ionospheric indices [34]. Of all the stations, the best fit of the model and experimental values of foF2 were obtained for the Salekhard station (66.5 °N, 66.5 °E).

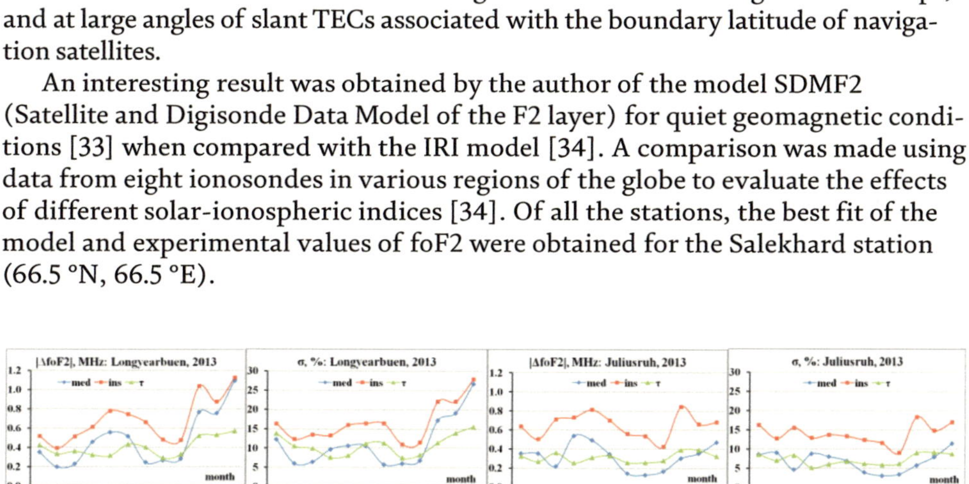

Figure 10.
Comparison of the accuracy of the model for the high-latitude station Longyearbuen and the reference mid-latitude Juliusruh.

	Longyearbuen			Juliusruh		
	med	**ins**	**τ**	**med**	**ins**	**τ**
2011	0.42, 11.90	0.64, 17.73	0.38, 11.73	0.44, 10.35	0.65, 14.96	0.28, 6.94
2012	0.36, 9.82	0.62, 17.18	0.43, 12.70	0.76, 17.95	0.99, 23.87	0.38, 9.13
2013	0.48, 11.40	0.69, 16.54	0.40, 10.71	0.32, 6.83	0.65, 13.98	0.31, 7.29
2014	0.70, 13.72	0.85, 17.34	0.42, 9.64	0.62, 11.47	0.83, 15.65	0.30, 6.22

Table 3.
Comparison of the accuracy of the IRI model from the data of high-latitude and mid-latitude stations.

The most important is the use of the model to specify propagation conditions on oblique paths. In [35], the calculated and experimental ionograms of oblique sounding were compared on 17 high-latitude HF paths of AARI during the period of quiet conditions on February 13–14, 2014 using the IRI-2012 model. It is shown that the IRI model without correction based on current diagnostic data, even under quiet conditions, underestimates the experimental values. Correction according to the current diagnostic data allows us to substantially approach the calculated values to the experimental ones. As the current diagnostics data, the critical frequencies measured by ionosondes located near the paths or calculated using the experimental values of TEC were used. It was obtained that the relative error of the initial IRI model in obtaining the values of the maximum useable frequency MUF averaged over all cases and the provided 23.6% for one hop was reduced by 4% when using the TEC and by 6% when using foF2. Analysis of experimental data showed that on high-latitude paths, there are a number of unpredictable features that occur even in quiet conditions. These include TID, M and N-modes, lateral modes, triplets and diffusivity. Below, the results of additional comparisons, including for disturbed days on September 6–8, 2017 are presented for ionograms of two oblique paths Cyprus-Lovozero (path length 3600 km) and Gorkovskaya-Lovozero (path length 900 km). It should be noted that the IRI model, like any statistical model, provides median (mean) values, so the model values should be compared with the experimental medians of the parameters. However, very often, model values are used as daily values, that is, instant, for the lack of others. In this case, the error can be large, especially during disturbances. In this case, such disturbance was observed from September 6 to 8, 2017. This fact is illustrated in **Figures 11** and **12**.

For this case, **Table 4** gives mean monthly (Δ), absolute (|Δ|) deviations and estimates of absolute and relative errors.

Relative errors lie in the range of 6–10%. Of course, this is a very small statistic, but it coincides with the results of [36] and allows us to confirm on the new data an important conclusion about the possibility of using the IRI model in high latitudes. **Figure 12** shows the difference in the disturbed values of foF2 from the medians. We see a significant positive disturbance in foF2 on September 7 and a negative perturbation in the behavior of both parameters on September 8. As it was noted earlier, a feature of the IRI model is its ability to be adapted to the current diagnostic data. This plays a big role in the calculation of oblique ionograms, which is illustrated in **Figure 13**. This figure shows the experimental MOF values for one and two hops, as well as the experimental value of foF2 in the center of the path. The black dots show the Dst index reduced in five times in absolute value. The right graph shows the values calculated by the numerical method. Red triangles refer to the experiment, black dots describe the curve for the original IRI model, and blue circles represent the results for model adaptation.

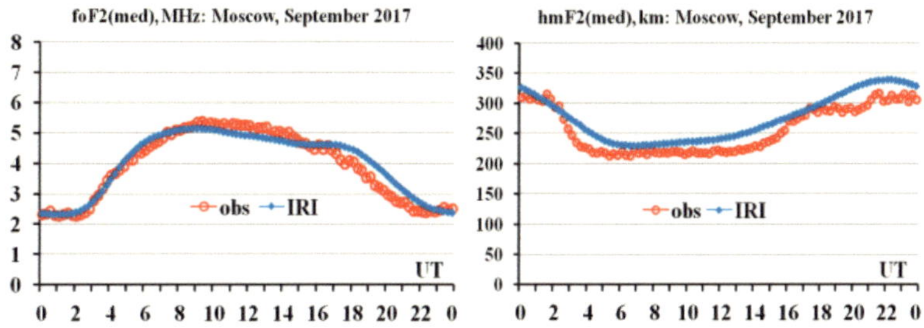

Figure 11.
Comparison of the model foF2 and hmF2 with experimental medians.

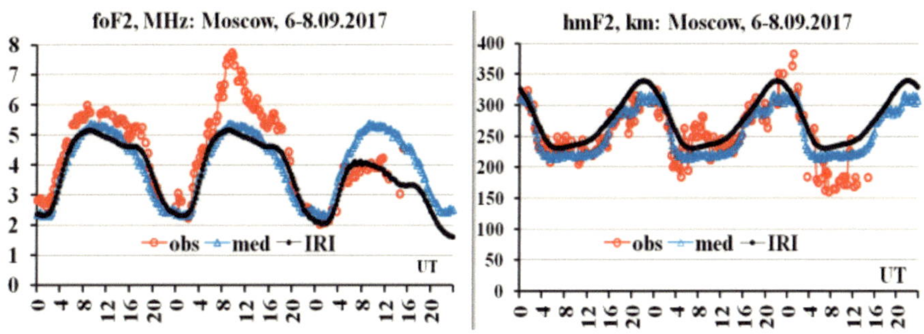

Figure 12.
Comparison of the model and experimental values of the parameters foF2 and hmF2 during the disturbed period September 6–8, 2017.

	Δ	|Δ|	RMS	RMS (%)
foF2	0.064 MHz	0.22 MHz	0.25 MHz	6.24
hmF2	18.9 km	19.5 km	26.7 km	10.4

Table 4.
Quantitative estimates of the correspondence between model and experimental medians.

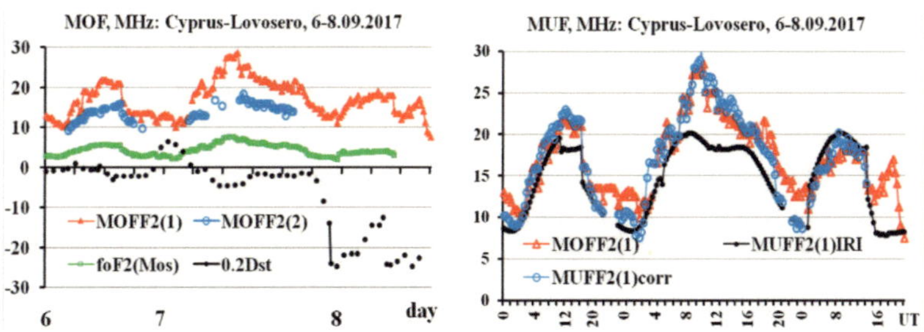

Figure 13.
Measured and model values of MUF from September 6 to 8 on the path Cyprus-Lovozero.

This allows us to conclude that the comparison of the parameters of HF propagation testifies that the model can be used in the high latitude region as yet, especially at adaptation to the data of the current diagnostics.

4. Areas of application of the proposed principle

Since the main area of application of the proposed principle is the polar region, it is necessary to point to an important result of [37]. It is shown that when recalculating a slant TEC into a vertical one, the traditional assumption of the ionosphere in the form of a thin layer is admissible. The height of this layer is associated with the position of the sub-ionospheric point. The calculated positions of this point are shown in **Figure 14**. The vertical axis represents the maximum latitude of the sub-ionospheric points in degrees, along the horizontal axis the latitude of the receiver in degrees is postponed.

This indicates that in high latitudes, there is no restriction on the visibility of the satellite, that is, on sounding the high-latitude ionosphere according to GNSS signals, and allows solving such tasks as tracking the position of high-latitude structures: mid latitude ionospheric trough (MIT) and auroral oval. The situation with the study of these structures is as follows. The trough generally consists of three parts: an equator wall, a trough minimum and a pole wall. Because of the large electron density, gradient on either side of the trough affects radio wave propagation, the exact position of the trough is very important for solving some problems, such as trans-ionospheric communication and navigation [38]. The main regularities of the behavior are obtained, and the MIT model is developed using foF2 [12] for night winter conditions—the period of the most probable occurrence of a trough. The corresponding behavior of the TEC shows that TEC can be used to identify the position of the dip [39]. It is shown that TEC always shows the presence of the trough.

The auroral oval in the high-latitude ionosphere is the boundary of the polar cap and is defined as the region of the ionosphere, which is the projection of the plasma layer and cusp along the lines of force of the geomagnetic field. The position of the auroral oval zone is projected onto the boundary of the outer radiation belt of the Earth. During magnetic storms, it shifts toward the mid-latitude, following a shift in the outer radiation belt. The displacement is almost linear and can reach 10 degrees with the growth of the Kp index to 5. With greater growth of Kp, the auroral oval boundary can jump discontinuously in the middle latitudes due to a significant distortion in the structure of the magnetosphere. In the auroral oval zone, the frequency of failures and refusals of radio communication and navigation equipment increases [37], so it is important to monitor its position. The possibility of such tracking not only in the meridional but also in the longitude directions is shown in [40].

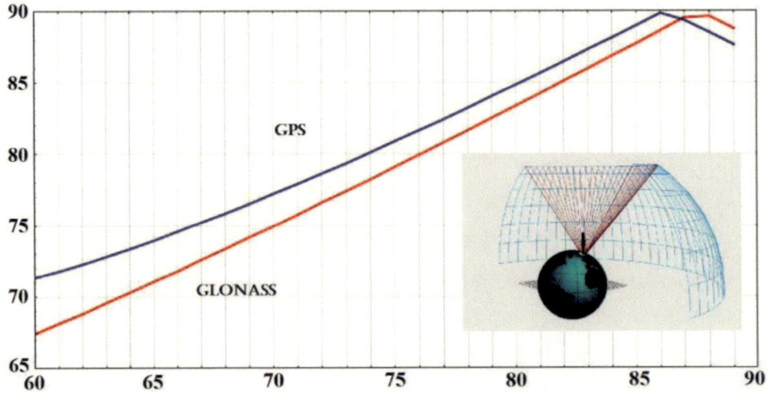

Figure 14.
Position of the maximum latitude of the sub-ionospheric point, depending on the latitude of the receiver.

An additional advantage of using the GLONASS platform for implementing the RTS method is associated with the ability to determine the N(h) profile from the initial height of the ionosphere to the height of the platform. This is due to the possibility of determining the critical frequency that can be used to adapt the model. The simulation results are shown in **Figure 15** for conditions of February 21, 2014, UT = 12, the satellite's altitude is 20,000 km, the satellite's latitude is 68.96°, longitude is 33.0°, the transmitter's latitude is 68.56°, longitude is 33.08°. The left panel of **Figure 15** shows the N(h)-profile as Fn(h)-curve (N is proportional to Fn^2), the middle—trajectories of waves, and the right panel shows the transionogram.

It can be seen that foF2 can be determined with an accuracy of 0.25 MHz. In experiments, it is possible to use the basic advantage of the model IRI-Plas, namely, its adaptation to measured value of TEC, which allows modifying plasmaspheric part of N(h)-profiles.

One more important possible application of the RTS method can be an installation of the receiver on low-flying mini satellites. The launch of such satellites is carried out constantly. It is possible to point out the results presented at the conference [41]. The micro-satellite "Chibis-M" was launched into a near-Earth orbit with parameters close to those of the ISS (513 km altitude, inclination 51°) on January 25, 2012 as a passing load on the cargo spacecraft Progress. It successfully operated for more than 2.5 years. The Vernov satellite was launched on July 8, 2014 to a solar-synchronous orbit with a small eccentricity—the height of the pericenter is 640 km, the altitude of the apocentre is 830 km, the inclination is 98.4°, and the period of revolution is 100 min. The Lomonosov satellite was launched on April 28, 2016 also to the solar synchronous orbit, circular with a height of 490 km, an inclination of 98.4° and a period of 90 min. The realization of these microsatellites has shown that with their help, it is possible to successfully carry out electromagnetic monitoring of the surrounding space environment. A new space project of the Moscow State University is being discussed (Lomonosov "Universal-SOKRAT") to create a grouping of satellites for real-time monitoring in near-Earth space. Methodological aspects of the spatiotemporal resolution of the plasma-wave parameters of the ionosphere with the help of two copies of Trabant MC (2020–2024), simultaneously deduced into an orbit with a height of ~500 km are considered. They intended for investigation of: (a) the mechanisms of occurrence and dynamics of ionospheric inhomogeneities of different scale depending on the active processes on the Sun and on Earth; (b) regularities of changes in plasma-wave and electromagnetic parameters in the ionosphere of natural and technogenic character in a wide dynamic and frequency ranges; (c) applied aspects, consisting in conducting diagnostics of ionospheric manifestations of space weather.

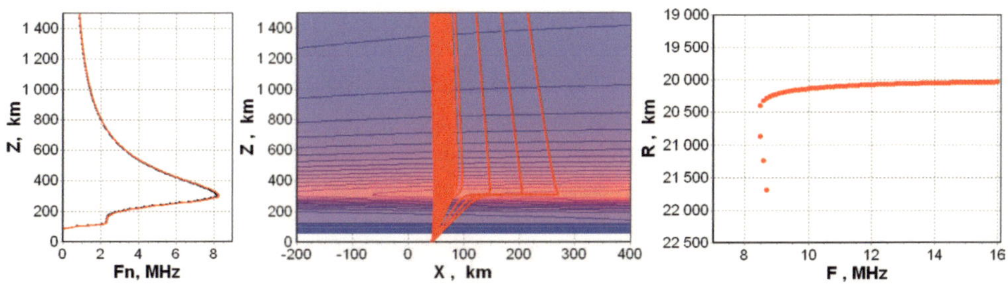

Figure 15.
Illustration of the possibility to obtain foF2.

5. Conclusions

The long-term use of satellite sounding data has made it possible to obtain the most important knowledge about near-Earth space. Climatological models of ionospheric parameters have been developed, which have found the widest application in various technological systems. However, at the present stage, it is necessary to provide operational support for these systems with ionospheric information. Here, it is necessary to look for ways of modifying different methods. In the ionospheric models, it is an adaptation to the parameters of the current diagnosis and the use of the total electron content for this diagnosis. A modification is also required in transionospheric sounding methods. It is shown that the main role can be played by the method of reverse transionospheric sounding, combined with the measurement of TEC. The installation of an on-board ionosonde receiver on the GLONASS platform helps to solve much problems. First, there is no need to develop a new platform for the polar high-apogee experiment, and second, the installation of the ionosonde receiver does not violate the electromagnetic compatibility requirements, the "timing" of the ionosonde to the exact time and the ephemerides of the satellite is simplified, and the transionogramms of the RTS can be transmitted through the service channels, besides, sizes of the reception aerial decrease at restriction of a frequency range [21].

Acknowledgements

This work was supported by grant under the state task N3.9696.2017/8.9 from Ministry of Education and Science of Russia.

Conflict of interest

There is no conflict of interest.

Author details

Igor Ivanov[1], Olga Maltseva[1*], Vladimir Sotskii[1], Alexandr Tertyshnikov[2] and Gennadii Zhbankov[1]

1 Institute for Physics, Southern Federal University, Rostov-on-Don, Russia

2 Institute of Applied Geophysics, Moscow, Russia

*Address all correspondence to: mai@ip.rsu.ru

References

[1] Goodman JM. Operational communication systems and relationships to the ionosphere and space weather. Advances in Space Research. 2005;**36**:2241-2252. DOI: 10.1016/j.asr.2003.05.063

[2] Florida CD. The development of a series of ionosphere satellites. Proceedings of the IEEE. 1969;**57**(6):867-875. DOI: 10.1109/PROC.1969.7132

[3] Givishvili GV et al. Transionospheric radiosonde based on the onboard ionosonde "LAERT" and ground ionosonde "PARUS-A" (in Russian). Heliogeophysical Research. 2015;**12**: 21-28. Available: http://vestnik.geospace.ru/index.php?id=301

[4] Danilkin NP. Transionospheric radiosounding (review). Geomagnetism and Aeronomy. 2017;**57**(5):501-511. DOI: 10.1134/S0016793217050048

[5] Vasiliev GV, Ivanov II, Kovalev SV. Device for the study of the ionosphere. Certificate of authorship. USSR № 1. 1986;340-383

[6] Denisenko PF, Sotskii VV. Specificfeaturesof inverse problems of vertical radio sounding of the ionosphere (review). News of Higher Educational Establishments. North-Caucasian Region. 1987;**2**:59-71

[7] Ivanov II. Synchronization of onboard and ground-based ionosondes in systematic sounding of the ionosphere. Physical Bases of Instrument Making. 2012;**1**:101-111

[8] Schmerling ER, Langille RC, guest editors. Special Issue on Topside Sounding and the Ionosphere. In: Proceedings of the IEEE. 1969;**57**(6)

[9] Jackson JE. Results-From-Alouette-1-Explorer-20-Alouette-2-And Explorer-31. July 1988. Available from: https://ru.scribd.com/document/48695743/Results-From-Alouette-1-Explorer-20-Alouette-2-And-Explorer-31

[10] Pulinets SA, Depuev VH, Karpachev AT, Radicella SM, Danilkin NP. Recent advances in topside profile modeling. Advances in Space Research. 2002;**29**(6):815-823

[11] Bilitza D, Reinisch BW, Radicella SM, Pulinets S, Gulyaeva T, Triskova L. Improvements of the international reference ionosphere model for the topside electron density profile. Radio Science. 2006;**41**:RS5S15. DOI: 10.1029/2005RS003370

[12] Karpachev AT, Klimenko MV, Klimenko VV, Pustovalova LV. Empirical model of the main ionospheric trough for the nighttime winter conditions. The Journal of Atmospheric and Solar-Terrestrial Physics. 2016;**146**:149-159. DOI: 10.1016/j.jastp.2016.05.008

[13] Benson RF, Osherovich VA. Application of ionospheric topside-sounding results to magnetospheric physics and astrophysics. Radio Science. 2004;**39**:RS1S28. DOI: 10.1029/2002RS002834

[14] Benson RF, Fainberg J, Osherovich VA, Truhlik V, Wang Y, Bilitza D, Fung SF. High-latitude topside ionospheric vertical electron-density-profile changes in response to large magnetic storms. Available from: https://ntrs.nasa.gov/search.jsp?R=201600058412018-06-29T13:55:48+00:00Z

[15] Givishvili GV, Danilkin NP, Zhbankov GA, Krasheninnikov IV. Possibilities of radio sounding of the ionosphere in the decameter range on board a geostationary satellite. Geomagnetism and Aeronomy. 2012;**52**(4):491-496. DOI: 10.1134/S0016793212040032

[16] Nava B, Coisson P, Radicella SM. A new version of the NeQuick ionosphere electron density model. The Journal of Atmospheric and Solar-Terrestrial Physics. 2008;**70**:1856-1862. DOI: 10.1016/j.jastp.2008.01.015

[17] Givishvili GV. MFRT method—The basis for remote monitoring the ionosphere of the RF polar zone in operative mode. Heliogeophysical Research. 2016;**14**:69-81. Available from: http://vestnik.geospace.ru/index.php?id=435

[18] Avdyushin SI, Danilkin NP, Ivanov AI, Ipatov EB, Kushnerevsky YV, Lukin DS, Migulin VV. Transionospheric sounding at the boundary of radio-transparency of the ionosphere. Geomagnetism and Aeronomy. 1983;**23**(4):567-572

[19] Danilkin NP, Zhbankov GA, Zhuravlev SV, Kotonayeva NG, Lapshin VB, Romanov IV. Monitoring of the ionosphere in the Arctic based on satellite ionosondes (in Russian). Heliogeophysical Research. 2016;**14**: 31-35. Available from: http://vestnik.geospace.ru/index.php?id=381

[20] Danilkin NP, Zhuravlev SV, Kotonaeva NG, Kuraev MA, Anishin MM. Modeling an experiment on radio sounding of the ionosphere from the artificial earth satellite Kosmos 1809 in the presence of vertical electron concentration inhomogeneities in the Arctic region. Geomagnetism and Aeronomy. 2012;**52**(2):229-234

[21] Ivanov II. Ionospheric monito ring in the arctic at reverse transionospheric sounding. In: Radiation and Scattering of Electromagnetic Waves (RSEMW); 26-30 June 2017; Divnomorskoe, Russia. 2017. pp. 118-120. DOI: 10.1109/RSEMW.2017.8103582. Available from: http://ieeexplore.ieee.org/document/8103582/

[22] Krasheninnikov IV. Analysis of the types of probing signals in the problem of ionospheric sounding and criteria for the efficiency of the use of space systems for conducting transionospheric monitoring in the Arctic (in Russian). Heliogeophysical Research. 2016;**14**: 53-62. Available from: http://vestnik.geospace.ru/index.php?id=383

[23] Tse D, Wiswanath P. Fundamertals of Wireless Communication. Cambridge, England: Cambridge University Press; 2005. 564 p

[24] Gulyaeva TL, Bilitza D. Towards ISO standard Earth ionosphere and plasmasphere model. In: Larsen RJ, editor. New Developments in the Standard Model. USA: NOVA Publishers; 2011. pp. 11-64

[25] Bilitza D, Altadill D, Truhlik V, Shubin V, Galkin I, Reinisch B, Huang X. International reference ionosphere 2016: From ionospheric climate to real-time weather predictions. Space Weather. 2017;**15**:418-429. DOI: 10.1002/2016SW001593

[26] Gulyaeva TL. Storm time behavior of topside scale height inferred from the ionosphere-plasmosphere model driven by the F2 layer peak and GPS-TEC observations. Advances in Space Research. 2011;**47**:913-920. DOI: 10.1016/j.asr.2010.10.025

[27] Maltseva OA, Zhbankov GA, Mozhaeva NS. Advantages of the new model of IRI (IRI-Plas) to study ionospheric environment. Advances in Radio Science. 2013;**11**:907-911. DOI: 10.5194/ars-11-307-2013

[28] Nickisch LJ. Practical applications of Haselgrove's equations for HF systems. Radio Science Bulletin. 2008;**325**:36-48. DOI: 10.23919/URSIRSB.2008.7909584

[29] Denisenko PF, Sotsky VV. Diagnosis of quasi-wave disturbances in near-earth plasma by modified SSA

method. In: Radiation and Scattering of Electromagnetic Waves (RSEMW); 26-30 June 2017; Divnomorskoe, Russia. 2017. pp. 17-20. DOI: 10.1109/RSEMW.2017.8103550. http://ieeexplore.ieee.org/document/8103550/

[30] Golyandina NE. A Method "Caterpillar"-*SSA*: The Analysis of Time Rows: The Manual. Stain Petersburg: BBM; 2004. 76 p

[31] Denisenko PF, Khomyakov AA. Monitoring large-scale moving ionospheric disturbances according to satellite sounding. Electromagnetic Waves and Electronic Systems. 2014;**19**(9):22-25

[32] Aitchison GJ, Weekes K. Some deductions of ionospheric information from the observations of emissions from satellite 1957a2-I. Journal of Atmospheric and Terrestrial Physics. 1959;**14**:236-243

[33] Shubin VN. Global empirical model of critical frequency of the ionospheric F2-layer for quiet geomagnetic conditions. Geomagnetism and Aeronomy. 2017;**57**(4):450-462. DOI: 10.1134/S0016793217040181

[34] Shubin VN. Comparison of solar-ionospheric indices for the foF2 modeling. In: Proceedings of the 2ndURSI AT-RASC; 28 May–1 June 2018; Gran Canaria

[35] Blagoveshchensky DV, Maltseva OA, Anishin MM, Rogov DD, Sergeeva MA. Modeling of HF propagation at high latitudes on the basis of IRI. Advances in Space Research. 2016;**57**:821-834. DOI: 10.1016/j.asr.2015.11.029

[36] Maltseva OA, Mozhaeva NS, Nikitenko TV. Comparison of model and experimental ionospheric parameters in the auroral zone. Advances in Space Research. 2013;**51**(4):599-609. DOI: 10.1016/j.asr.2012.04.009

[37] Kovalev DS, Tertyshnikov AV, Chukin VV, Glukhov YV. Experiments on ionospheric research on board of the arctic floating university. Scholarly Notes. 2016;**41**:156-164

[38] Yang N, Le H, Liu L. Statistical analysis of ionospheric mid-latitude trough over the Northern Hemisphere derived from GPS total electron content data. Earth, Planets and Space. 2015;**67**:196. DOI: 10.1186/s40623-015-0365-1

[39] Maltseva O. Verification of Ionospheric Models by TEC and Satellite Measurements ICTRS'17; November 6-7, 2017; Delft, Netherlands© 2017 Association for Computing Machinery. ACM. ISBN 978-1-4503-6364-8/17/11...$15.00 https://doi.org/10.1145/3152808.3152817

[40] Tertyshnikov AV. The manifestation of the auroral oval over Antarctica in the characteristics of GNSS signals 17.01.2015 near the station "Vostok". Heliogeophysical Research. 2016;**14**: 31-35. Available from: http://vestnik.geospace.ru/index.php?id=459

[41] Abstracts of 13th annual conference "Physics of plasma in solar system"; February, 12-16th 2018; Moscow, ICR. 383 p

Section 4

High Resolution Satellite Data Application

Chapter 5

High-Resolution Satellite Imagery Classification for Urban Form Detection

Juan Manuel Núñez, Sandra Medina, Gerardo Ávila and Jorge Montejano

Abstract

Mapping urban form at regional and local scales is a crucial task for discerning the influence of urban expansion upon the ecosystem and the surrounding environment. Remotely sensed imagery is ideally used to monitor and detect urban areas that occur frequently as a consequence of incessant urbanization. It is a lengthy process to convert satellite imagery into urban form map using the existing methods of manual interpretation and parametric image classification digitally. In this work, classification techniques of high-resolution satellite imagery were used to map 50 selected cities of study of the National Urban System in Mexico, during 2015–2016. In order to process the information, 140 RapidEye Ortho Tile multispectral satellite imageries with a pixel size of 5 m were downloaded, divided into 5 × 5 km tiles and then 639 tiles were generated. In each (imagery or tile), classification methods were tested, such as: artificial neural networks (RNA), support vector machines (MSV), decision trees (AD), and maximum likelihood (MV); after tests, urban and non-urban categories were obtained. The result is validated with an accuracy method that follows a stratified random sampling of 16 points for each tile. It is expected that these results can be used in the construction of spatial metrics that explain the differences in the Mexican urban areas.

Keywords: urban form, remote sensing, high-resolution satellite imagery, advanced classification methods, GIS integration

1. Introduction

Urbanization, as a process that manifests itself through the concentration of population in cities, is considered one of the most powerful and visible anthropogenic forces on the planet. Its influence is manifested on topics ranging from environmental changes on a global, regional, and local scale [1, 2], socioeconomic problems [3] to urban planning [4]. Thereby, several investigations use maps of urban areas to assess the influence of urbanization on natural and human environments and to estimate some important aspects of urbanization, such as its composition [5], size, scale, and form [6].

The urban form is the most visible result of the economic, social, cultural, and environmental driving forces of urban development [1]. Therefore, it is a spatial reflection of different processes across the evolution of a city and its characterization is a valuable source of information for urban planning. Ultimately, urban form is the result of the symbiotic interactions of infrastructures, people, and economic activities in a city that is constantly evolving in response to social, environmental, economic, and technological development [7].

In the cities, urban form is materialized by the heterogeneous physical alignment and characteristics of buildings, streets, and open spaces at different levels of spatial resolution. This high heterogeneity of materials and urban objects in terms of size, forms, and urban fabric morphology of the cities can be detected through the use of remote sensing imagery. This type of research provides very important information in relation to urban issues on planning, housing, health, transportation, and economic policies; especially for regions in developing countries that are less documented.

Most of the research efforts have been made for mapping urban landscapes at various scales and on the spatial resolution requirements of such mapping [8]. Different remote sensing techniques have already shown their value in mapping urban areas with different spatial, geometric, spectral, and temporal resolutions for different purposes. Therefore, the selection of an appropriate estimation method based on remotely sensed data characteristics is important.

Traditional remote sensing literature review suggests that major approaches include pixel-based image classification [9, 10], spectral index [11, 12], object-oriented algorithms [13, 14], and machine learning like artificial neural networks [15] and decision tree classification algorithm [16]. Techniques, such as data/image fusion, have also been explored [17]. Recent research has used high and very high spatial resolution remote sensing imagery to quantitatively describe the spatial structure of urban environments and characterize patterns of urban morphology [18].

Remote sensing approach compared with traditional methods for mapping the urban form provides certain advantages due to its convenience, efficiency, and coverage [19]. For this reason, the study of the detection of the urban form and its corresponding derived attributes through different types of satellite images is becoming of more interest [16, 20–23].

Regardless of the satellite imagery classification method employed for urban form detection, they can be divided into two categories: supervised and unsupervised methods. Those results obtained by the first ones usually produce a greater reliability, nevertheless they require more processing steps for the construction of training data.

For the supervised methods, the classifiers based on support vector machines (SVM) are very popular due to their good performance and robustness [24, 25]. Additionally, the methods based on the artificial neural networks (ANN) are also widely used for the classification of urban areas [26]. For example, Dridi et al. [27] combine multiple SVM for the mapping of urban extensions in the city of Algeria and compare them with ANN to support the experimental analysis to monitoring the spatiotemporal phenomenon of urban sprawl. Other supervised classification methods, such as decision tree (DT), regression model (RM), and maximum likelihood (ML), can also provide plausible results in the mapping of urban areas [28].

In this work, we evaluated four supervised classification methods (SVM, ANN, DT, and ML) using satellite images of earth observation, to integrate with a GIS approach the mapping of the urban form in 50 Mexican cities. The rest of this document is organized as follows: in Section 2, the context of the cities selected for the test and the dataset used are briefly presented; in Section 3, it is described the methodology with the proposed classification strategy for urban mapping that includes the preprocessing of RapidEye images, the collection of training samples, the classification methods evaluating the validation strategy, and the postprocessing GIS approach. The experimental results obtained and their discussions are presented in Section 4. Finally, the conclusions of the work are expressed in Section 5.

2. Context

2.1 Study area

In Mexico, urbanization has been associated with increased prosperity and improvements in quality of life. Urban areas, lead in expanding coverage of basic and social services, also offer better access to other services and amenities, including health care and education. Moreover, Mexico's growing middle class and declining inequality in recent decades seem to be definitely urban phenomena [29].

There have been important changes on the spatial form of Mexican cities over the past 30 years: most notably urban growth is characterized as distant, dispersed, and disconnected. Between 1980 and 2010, the built-up area of Mexican cities expanded on average by a factor of seven and the urbanized area of the 11 biggest metropolitan areas with more than 1 million inhabitants in 2010 has even grown by a factor of nine (SEDESOL 2012). This rapid spatial transformation of most Mexican cities presents important challenges for their potential to promote green and inclusive growth. To solve these problems, different initiatives have made significant efforts to put in place measurement systems and to broaden information about urban dynamics.

An ambitious national initiative, the National Urban System (NUS) is a unified platform to support decision-making for urban and housing policies. The NUS, launched by Mexican federal agencies in 2012, exemplifies a significant effort to broaden information and understanding about urban dynamics and has been recognized as innovative among Latin American urban initiatives. This system is a reference to analyze spatial patterns of Mexican cities, their causes, and their impact and to provide an analytical basis to understand urban phenomenon.

The National Population Council (Consejo Nacional de Población, CONAPO) and the Secretariat of Social Development (Secretaria de Desarrollo Social, SEDESOL) put together the NUS on the basis of data from the Population and Housing Census (2010) with the objective of creating a system to support strategic planning and decision-making in urban areas and to provide all sectors (state governments, municipalities, academia, private sector, and general users) with integrated metropolitan and urban information on demographic and socioeconomic variables. The NUS comprises 384 cities with over 15,000 inhabitants each, out of which 59 are metropolitan areas, 78 conurbations (suburban centers), and 247 urban centers. About 81.2 million people or 72.3% of the country's population live in these 384 cities.

The study area corresponds to a 50 cities sample of the NUS that include three types of cities, classified on the basis of geographical delimitations defined by the NUS (**Figure 1**).

These 50 urban areas include:

i. *12 metropolitan areas* defined as a group of municipalities that share a central city and are highly integrated with more than 250,000 residents: (1) Aguascalientes, (2) Monclova-Frontera, (3) Juárez, (4) San Francisco del Rincón, (5) Moroleón-Uriangato, (6) Tula, (7) Tehuacán, (8) Rioverde Ciudad Fernández, (9) Nuevo Laredo, (10) Coatzacoalcos, (11) Tianguistenco, and (12) Teziutlán.

ii. *16 urban conurbations* that extend across more than one locality and have more than 15,000 residents: (13) Ensenada, (14) Campeche, (15) Manzanillo, (16) Tapachula de Córdova y Ordóñez, (17) Guanajuato, (18) Irapuato, (19) Chilpancingo de los Bravo, (20) Ciudad

Lázaro Cárdenas, (21) Uruapan, (22) Zitácuaro, (23) San Juan Bautista Tuxtepec, (24) Chetumal, (25) Ciudad Obregón, (26) Cárdenas, (27) Túxpam de Rodríguez Cano,and (28) Fresnillo.

iii. *22 urban centers* that have more than 15,000 residents and that do not extend beyond the boundaries of their locality: (29) La Paz, (30) Ciudad del Carmen, (31) Ciudad Acuña, (32) Comitán de Domínguez, (33) San Cristóbal de las Casas, (34) Cuauhtémoc, (35) Delicias, (36) Hidalgo del Parral, (37) Victoria de Durango, (38) Salamanca, (39) Iguala de la Independencia, (40) Ciudad Guzmán, (41) Lagos de Moreno, (42) Apatzingán, (43) San Juan del Río, (44) Ciudad Valles, (45) Los Mochis, (46) Culiacán Rosales, (47) Mazatlán, (48) Navojoa, (49) Heroica Nogales, and (50) Ciudad Victoria.

2.2 Materials

Urban areas were identified by looking at the layer of urban polygons of the geostatistical framework, version 5.0 of the National Institute of Statistics and Geography (Instituto Nacional de Estadística y Geografía, INEGI). Later, satellite images were obtained for the binary classification between urban and nonurban areas that covered the 50 study cities, for which 140 RapidEye images of the period 2015–2016 were acquired, through the Planet platform (www.planet.com).

The main characteristics of these images are: (a) spatial resolution of 5 m and covered area per image of 25 km^2; (b) 5-band spectral resolution (blue 440–510 nm, green 520–590 nm, red 630–685 nm, red edge 690–730 nm, and near-infrared 760–850 nm); (c) 12-bit radiometric resolution, and (d) Universal Transverse Mercator (UTM) and WGS84 Horizontal Datum.

Additionally, a digital elevation model (DEM) of the Mexican territory was downloaded to perform the radiometric and atmospheric corrections. Finally, for the collection of training samples, a Web Map Service (WMS) of a SPOT satellite

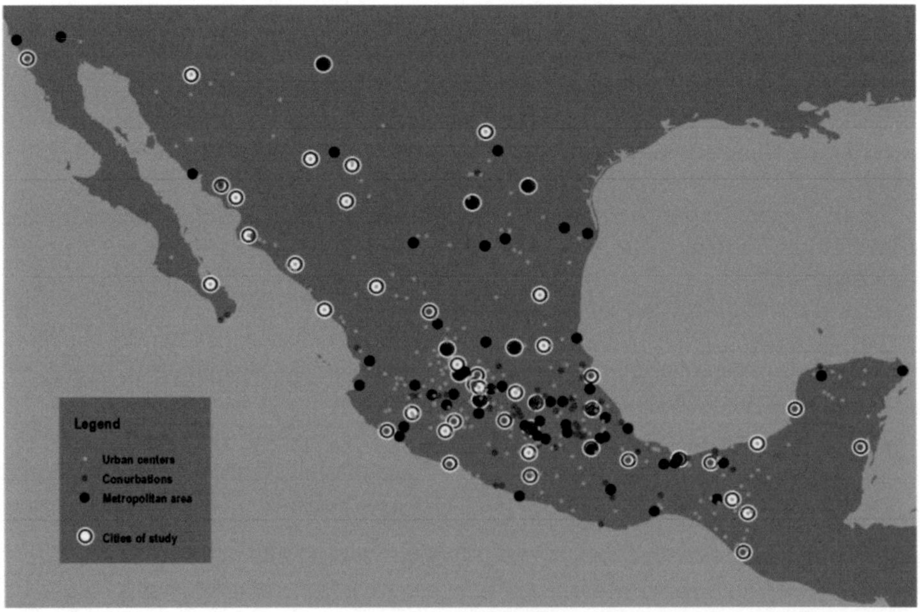

Figure 1.
Selected cities of study, National Urban System and classification of city types. Source: Own elaboration based on data from the secretariat of social development (Secretaría de Desarrollo Social, SEDESOL).

images mosaic provided by the Mexico Reception Station (Estación de Recepción México, ERMEX) was used, at a resolution of 1.5 m in true color.

3. Methodology

The methodology is split into five main steps as follows: strategy for satellite imagery download and preprocessing, training and validation sample selection, classification methods, GIS integration, and results evaluation.

3.1 Strategy for satellite imagery download and preprocessing

In the first step, the entire Mexican territory was divided into nonoverlapping 5 × 5 km blocks, with the purpose of selecting blocks that cover the mosaics of the images related to the urban areas selected. A total of 639 blocks were selected to cover the 50 urban areas. Then, 140 RapidEye Ortho Tile multispectral scenes were downloaded through the Planet platform (www.planet.com) to cover all cities within the project. The satellite images were selected for the period 2015–2016, obtaining a homogeneous selection of acquisition dates and conditions of zero or little cloudiness.

Radiometric and atmospheric corrections were conducted to retrieve surface reflectance values by means of the atmospheric and topographic corrections software (ATCOR3) implemented in the ENVI virtual IDL machine [30]. Finally, mosaics by blocks were prepared for each of the 50 cities.

3.2 Training and validation sample selection

To obtain training and validation samples, the generated blocks in the previous stage were used to cover the mosaics of the satellite imagery that corresponds to the selected. Training and validation data should be representative of the study area and of the classification scheme. Because urban is often a relatively rare class that covers only a small proportion of the landscape, spatial stratification with proportional class allocation (SpatialProp) was selected to be able to obtain high user's accuracy of urban class [31].

In the SpatialProp strategy, the sample size is allocated to each class proportional to the areal coverage in the reference set, with the constraint that each spatial stratum receives an equal total sample size. For example, if the urban and nonurban classes comprised 25 and 75% of the area of the entire region, respectively, the sample allocation in each spatial stratum would be 25% urban and 75% nonurban. According to Jin et al. [31] in each 5 × 5 km block, 16 random samples are assigned to the urban and nonurban strata proportional allocation. For example, in our hypothetical situation, nonurban occupies 75% of the area and urban occupies 25%. Given the total sample size of 16, 12 nonurban pixels and 4 urban pixels will be selected following the designs of SpatialProp.

For the 639 blocks employed for the 50 selected urban areas, 20,448 sampling and validation points were assigned. Later, each of the data points were verified with the related category based on the RapidEye mosaic and the Web Map Service (WMS) of a SPOT Image.

3.3 Classification methods

Machine-learning classification has become a major focus of the remote-sensing literature since it is generally able to model complex class signatures without making assumptions about the data distribution, i.e., it is nonparametric [25]. A wide range of studies have generally found that these methods tend to produce higher accuracy compared to traditional parametric classifiers, especially for complex data with a high-dimensional feature space [32, 33].

However, parametric maximum likelihood (ML) classifier method is the most commonly used remote-sensing classification method [34]. In this work, we evaluate the classification methods of artificial neural networks (ANN), support vector machines (SVM), decision tree (DT), and maximum likelihood (ML) for each city. For each of this classifier, we can measure the accuracy based on the use of an error matrix. Below, there is a brief description of each referred methods.

3.3.1 Artificial neural networks (ANN)

An artificial neural network is a massive parallel distributed processor made up of simple processing units, which has a natural propensity for storing experiential knowledge and can make it available for use [35]. The model is formed by artificial neurons that emulate biological neurons and the synaptic connections among them; it regulates them through the process of solving problem [36].

The network needs to be "trained" with a sufficiently large number of examples in order to be able to make the appropriate inferences. The procedure of training involves groups of input data together with the expected output data. Once the system of neurons has been trained, the network allows the processing of imprecise information, the generalization of known responses to new situations, and the prediction of outcomes. They are appropriate models for dealing with a large set of variables and their nonlinearity is convenient for the assessment of complex systems [37].

The links with the neurons located in the so-called hidden neuron layer take then different weights and are educated depending on the required output, thus they can model complex relationships among variables. The system requires feed-forward and backpropagation processes to allow the network to get trained [38]. The visualization of this stage is accomplished through error analysis. If the error becomes smaller and asymptotic, the network will be ready to receive new input data and to predict an output [37].

The ANN models used in this study are of the multilayer perceptron ANN type, a model in which all neurons are fully connected to adjacent layers while layers are not connected to each other at all [39, 40]. There are three types of layers in a typical multilayer perceptron network: input layer, hidden layer, and output layer. This architecture is shown in **Figure 2**. In each case, the training of the proposed network was performed with a backpropagation algorithm which is a supervised learning procedure [41].

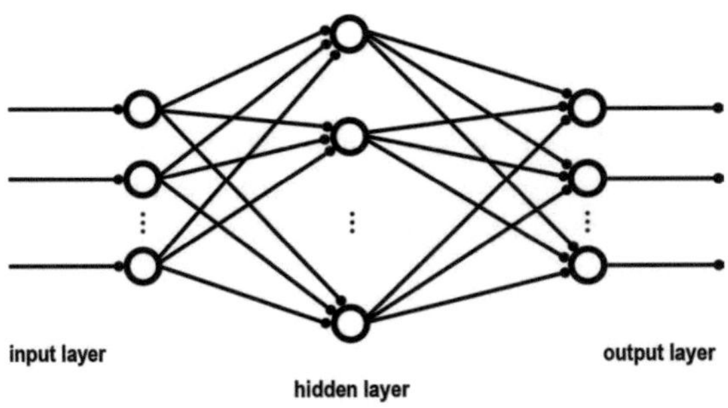

Figure 2.
Artificial neural networks classifier. Source: adapted from [39].

The main tasks of remote sensing data analysis in which the application of ANN standard backpropagation for supervised learning is reported are classification, more commonly land cover classification [42, 43], unmixing [44, 45], and retrieval of biophysical parameters of cover [46]. Other applications of ANNs are also reported in change detection, data fusion, forecasting, preprocessing, georeferencing, and object recognition.

3.3.2 Support vector machines (SVMs)

Support vector machines are a supervised nonparametric statistical learning technique that has no assumption made on the underlying data distribution [47]. Initially, the method is presented with a set of labeled data instances and the SVM training algorithm aims to find a hyperplane that separates the dataset into a discrete predefined number of classes in a fashion consistent with the training examples [48]. Where, optimal separation hyperplane term is used to refer to the decision boundary that minimizes misclassifications, obtained in the training step and learning refers to the iterative process of finding a classifier with optimal decision boundary to separate the training patterns (in potentially high-dimensional space) and then to separate simulation data under the same configurations (dimensions) [49].

In its simplest form, SVM are linear binary classifiers that assign a given test sample a class from one of the two possible labels [47]. **Figure 3** illustrates a simple scenario of a two-class separable classification problem in a two-dimensional input space where the solution for a typical two-dimensional case where the subset of points that lies on the margin (called support vectors) is the only one that defines the hyperplane of maximum margin.

An important generalization aspect of SVMs is that frequently not all the available training examples are used in the description and specification of the separating hyperplane. The subset of points that lie on the margin (called support vectors) is the only one that defines the hyperplane of maximum margin. If the two classes are not linearly separable, the SVM tries to find the hyperplane that maximizes the margin while, at the same time, minimizing a quantity proportional to the number of misclassification errors [50]. The tradeoff between margin and misclassification error is controlled by a user-defined constant [51]. SVM can also be extended to handle nonlinear decision surfaces. Boser et al. [52] propose a method of projecting

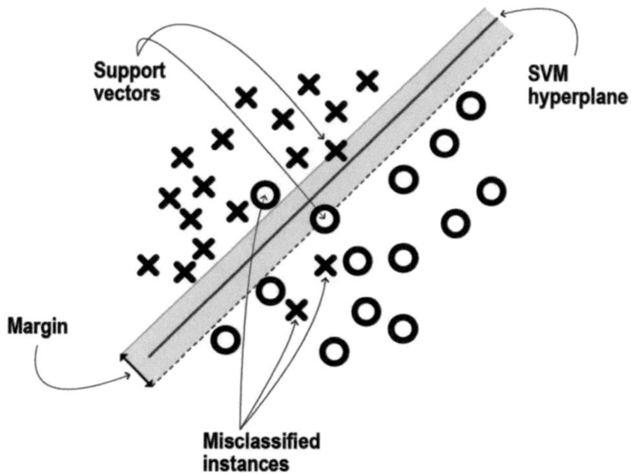

Figure 3.
Linear support vector machine classifier. Source: adapted from [47].

the input data onto a high-dimensional feature space using kernel functions and formulating a linear classification problem in that feature space [53].

In case of nonlinear classification, SVM can perform the classification by using various types of kernels which turn nonlinear boundaries to linear ones in the high-dimensional space to define optimal hyperplane [54]. In this study, four types of kernels (linear, polynomial, radial basis function, and sigmoid) were used for the SVM classification.

3.3.3 Decision tree (DT)

A decision tree is a flow chart like tree structure, defined as a classification procedure that recursively partitions a dataset into smaller subdivisions on the basis of a set of tests defined at each branch (or node) in the tree [55]. **Figure 4** illustrates a tree composed of a root node (formed from all of the data), a set of internal nodes (splits), and a set of terminal nodes (leaves). Each circle is a node at which tests (T) are applied recursively, in order to split the data into smaller groups. The labels (A, B, C) at each leaf node refer to the class label assigned to each observation.

In this framework, a DT classifier performs multistage classifications by using a series of binary decisions to place pixels into classes. Each decision divides the pixels in a set of images into two classes based on an expression. It is possible to divide each new class into two more classes based on another expression and defines as many decision nodes as needed. Decision trees have significant intuitive appeal because the classification structure is explicit and therefore easily interpretable since the results of the decisions are always classes. Furthermore, it is possible to use data from many different sources and files together to make a single DT classifier.

The construction of decision tree classifier does not require any domain knowledge of parameter setting, and therefore, is appropriate for satellite imagery classification [56]. The learning and classification steps of decision tree induction are simple and fast. In general, decision tree classifier has good accuracy. Decision tree induction algorithms have been used for classification in many applications areas, including remote sensing [57]. Decision trees have several advantages over traditional supervised classification procedures used in remote sensing such as l ISODATA clustering and maximum likelihood classifier algorithms [58]. In

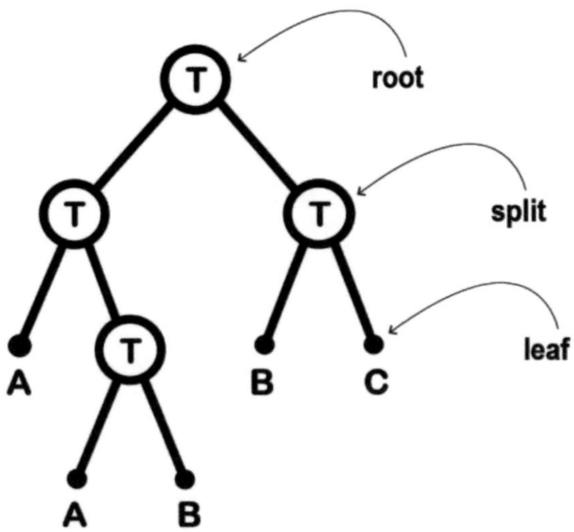

Figure 4.
Decision tree classifier. Source: adapted from [55].

particular, decision trees are strictly nonparametric and do not require assumptions regarding the distributions of the input data. In addition, they handle nonlinear relations between features and classes, they verify missing values and are capable of handling both numeric and categorical inputs in a natural manner [55].

3.3.4 Maximum likelihood (ML)

Into the classic remote sensing image classification techniques, maximum likelihood (ML) classifier, widely implemented in commercial image-processing software packages, is the most frequently method used to pixel-wise classification [34]. ML classifier assumes that the statistics for each class in each band is normally distributed and calculates the probability that a given pixel belongs to a specific class. Unless the algorithm selects a probability threshold, all pixels are classified. Each pixel is assigned to the class that has the highest probability, that is, the maximum likelihood [41].

Statistical techniques such as ML estimation usually assume that data distribution is known a priori [59]. The ML algorithm in remote sensing classification is parametric and depends on each class and is represented by a Gaussian probability density function, which is completely described by the mean vector and variance–covariance matrix using all available spectral bands, and if possible, ancillary information (**Figure 5**). The maximum likelihood classifier is based on an estimated probability density function for each of the reference classes under consideration, where the class statistics is obtained from the training data. Given these parameters, it is possible to compute the statistical likelihood of a pixel vector as a member of each spectral class [60].

The maximum likelihood classifier is simple and robust enough to accommodate modifications. With the advent of commercial high and very high spatial resolution sensor data, the ML classifier is appropriate for many urban applications [61]. In the context of the new generation of very high spatial resolution commercial satellite sensors, data from these sensors are high volume and they measure large spectral variations in urban land cover, so that in the absence of classifiers designed to deal with such data, simplicity in the maximum likelihood can accommodate large datasets, and the modifications outlined [62].

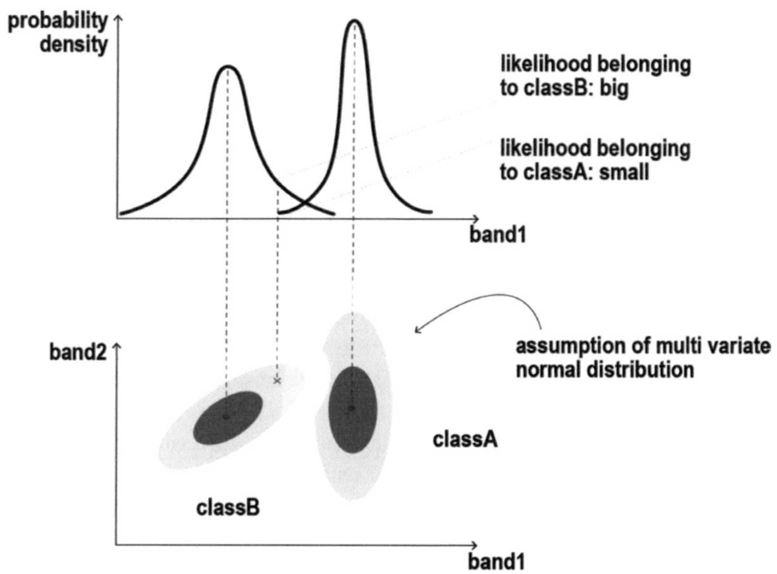

Figure 5.
Maximum likelihood classifier. Source: adapted from [59].

3.4 Validation strategy

In this step, the overall classification accuracies were determined from the error matrix by calculating the total percentage of pixels correctly classified for the classification methods of: (i) artificial neural networks (ANN); (ii) support vector machines (SVM) for linear (ML), polynomial (MP), radial basis function (MRBF), and sigmoid (MS) kernels; (iii) decision tree (DT); and (iv) maximum likelihood (ML). Since this assessment takes only the diagonal of the matrix into account, the Kappa coefficient, which is based on all the elements in the confusion matrix, was also calculated [63]. The overall accuracy and kappa values were determined using test datasets, obtained with the SpatialProp strategy for training and validation samples developed in Section 3.2.

With the approach of more advanced digital satellite remote sensing techniques, the necessity of performing an accuracy assessment has received renewed interest [64]. Accurate assessment or validation is an important step in the processing of remote sensing data. At present, the geographic information systems and remote sensing communities are becoming more interested on accurate topics. Technological developments in the area of data processing offer more and more possibilities. In this work, the collection of training samples collected from a Web Map Service (WMS) of a SPOT satellite images mosaic at a resolution of 1.5 m in true color is used. The data collected by this method are comparable to the field data employed to assess the accuracy of these remote sensing products.

3.5 GIS integration

The different nonparametric classifiers implemented in this work, such as an artificial neural network, decision tree, support vector machines, and the traditional maximum likelihood classifier, have their own strengths and limitations. For example, when sufficient training samples are available and the feature of land covers in a dataset is normally distributed, a maximum likelihood classifier may yield an accurate classification result. In contrast, when an image data are anomalously distributed, neural network and decision tree classifiers may demonstrate a better classification result [65, 66]. Some other times, machine-learning approaches provide a better classification result than ML, although some tradeoffs exist in classification accuracy, time consumption, and computing resources [67].

Previous research has indicated that the integration of two or more classifiers provides improved classification accuracy compared to the use of a single classifier [67–69]. A critical step is to develop suitable rules to combine the classification results from different classifiers. Some previous research has explored different techniques, such as a production rule, a sum rule, stacked regression methods, majority voting, and thresholds, to combine multiple classification results [69, 70].

In this step, we have employed a GIS approach to integrate the results of the ANN, SVM, DT, and ML classifiers to produce a better final map of urban form. Different urban mapping hybrid approaches have already been combined to achieve better results [71, 72]. In our approach, the matching results of two or more methods evaluated are combined by the superposition function with the results of the best evaluated method. Subsequently, through a selection of these attributes, the pixels of the urban and nonurban uses that were identified as the best results of the combination within a GIS environment are extracted. The resulting map was validated again, revealing that the most likely characteristics of urban and nonurban uses were present in the combined pixels. This integration GIS approach has allowed the improvement of the results of the urban area classification for the selected cities of study. We suggested that this integration approach can be economically and immediately implemented in a standard GIS software package to produce urban form maps with higher accuracy from satellite images of high spatial resolution for the Mexican National Urban System.

4. Results and discussion

In this study, four different supervised classification methods were integrated to map urban forms of 50 selected cities of study in the National Urban System in Mexico. Maximum likelihood classifier which is a conventional classification method and the advanced classification methods: artificial neural networks, decision tree, and support vector machines for linear (ML), polynomial (MP), RBF (MR), sigmoid (MS) kernels. We found that the artificial neural network classifier (overall accuracy of 92.2%) turned out to be the better single classification method. Support vector machine (overall accuracy of 89.8%) and maximum likelihood (overall accuracy of 89.2%) had similar results. Decision tree classification method (overall accuracy of 87.8%) was the lower classification method. The results we obtained were evaluated by the overall accuracy which is computed by dividing the total number of correct pixels (i.e., the sum of the major diagonal) by the total number of pixels in the error matrix. Overall accuracy for ANN, DT, selected SVM models, and ML classifiers is summarized in **Figure 6**.

After integrating the results obtained by city, using GIS approach, each evaluated method produces a result that has an impact on the spatial extent of the urban form, this is an important result. GIS approach showed an overall accuracy above the average of global reliabilities for each of the 50 selected cities of study; the average reliability for the methods evaluated in all the cities was 89.8%; when using GIS approach, this average reached 91.2%; this number is higher in 38 of the 50 cities evaluated. The approach used in this work has shown good results, although all the classifiers showed

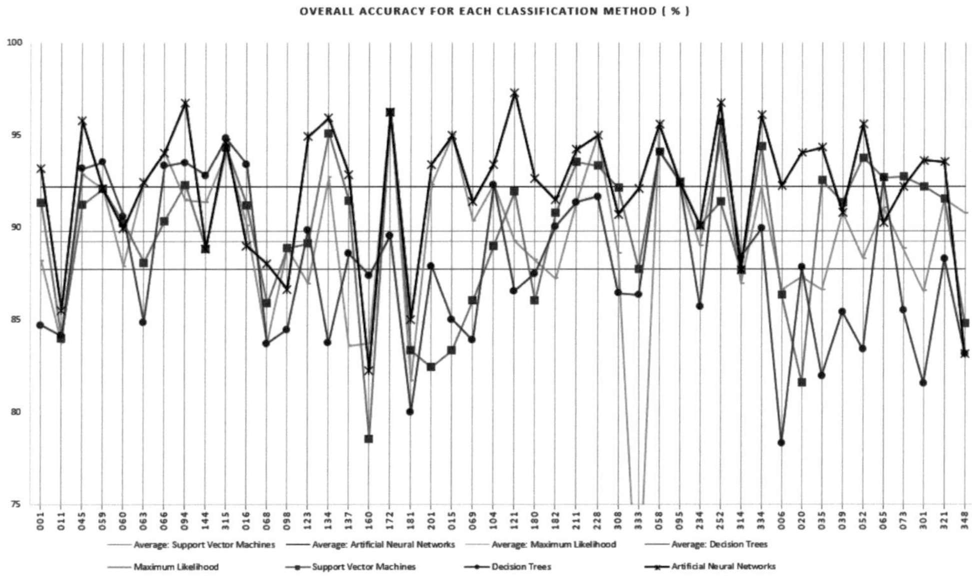

Figure 6.
Overall accuracy for ANN, DT, selected SVM models, and ML classifiers. Source: own elaboration.

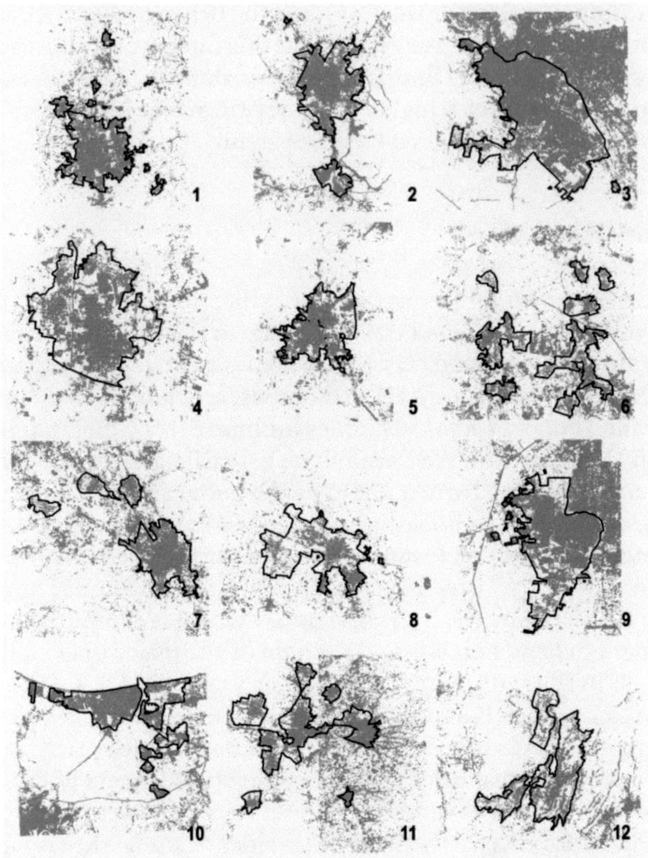

a

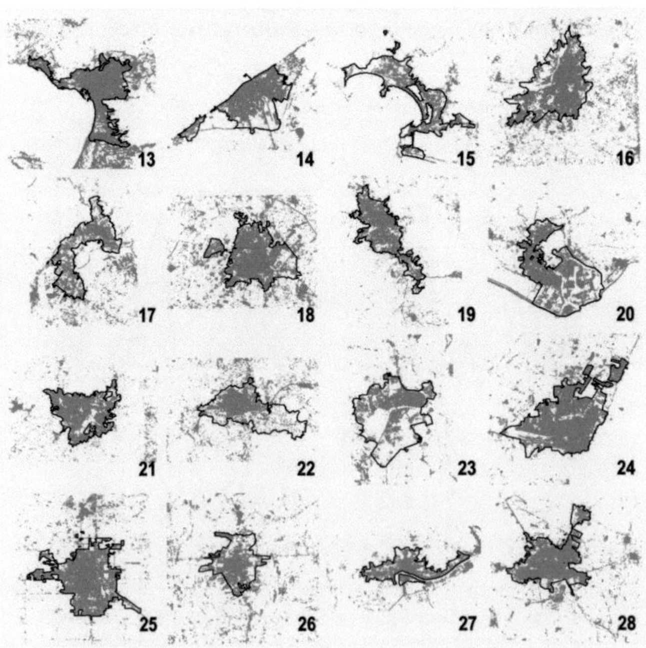

b

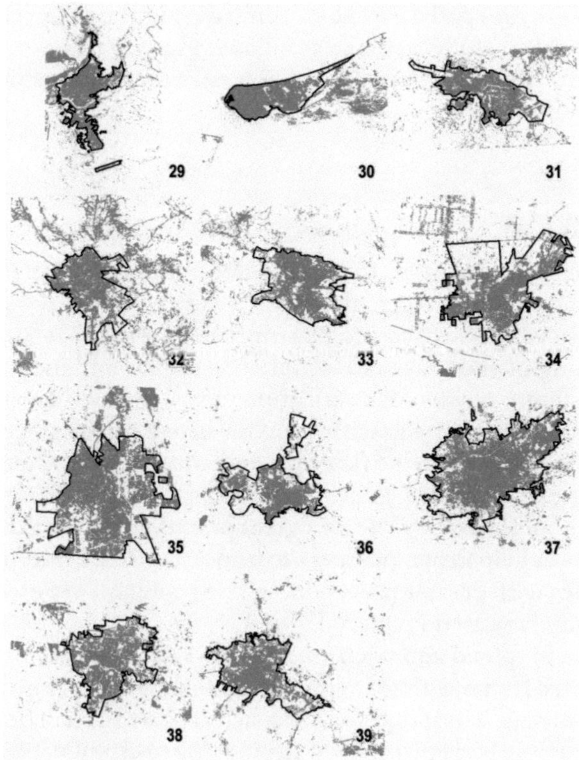

c-1

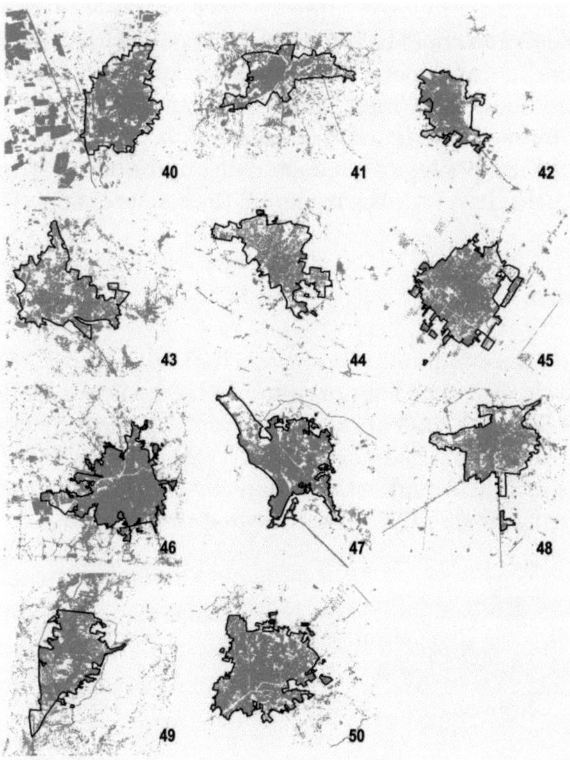

c-2

Figure 7.
(a) Metropolitan areas. Source: own elaboration. (b) Urban conurbations. Source: own elaboration. (c-1) Urban centers 29–38. Source: own elaboration. (c-2) Urban centers 40–50. Source: own elaboration.

very little differences in the spatial extent (within ±4%) of the urban class. The result for the 50 selected cities of study is shown as follows. **Figure 7a** shows the metropolitan areas, **Figure 7b** the urban conurbations, and **Figure 7c** the urban centers.

5. Conclusions

Information about urban form maping is essential for proper planning and to examine how the recent urban growth has affected the economic performance and livability of cities. This methodological approach offers a spatially explicit inputs for adjusting urban policy frameworks and instruments in ways that support sustainable spatial development and make cities more productive and inclusive.

In this work, different advance classification methods have been tested for the high-resolution satellite imagery classification for urban form detection. SVM method proved to be better for classification problems of two classes. Its major advantage is the less parameters to make it operational and reach high accuracy rates. The employed methodology shows a great potential for the urban form mapping, which could help urban planners to understand and interpret complex urban characteristics with greater precision, where problems are often cited about satellite-based remotely sensed imagery [73].

Furthermore, the proposed approach used to integrate results through GIS environment indicates a robust framework for addressing integrated classification problems in the field of remote sensing. This proposed approach allows to obtain better results when is used to integrate, on the basis that each of the integrated classification methods provides the best of its results to the benefit of a more accurate urban form classification.

Therefore, we believe this proposed approach has great practical value for several remote sensing problems and could be improved and applied to various urban applications in the near future. In this respect, this integration approach can be strengthened through the implementation of learning methods to manage the integration of the data and therefore obtain more and better reliable results. Finally, we are also interested in plainly analyzing the morphological characteristics of the urban form through the application of metrics that have, as primary input, the results obtained with this work.

Acknowledgements

The authors thank the anonymous reviewers for their comments and suggestions. We also thank the financial support granted by the Fondo Sectorial INEGI-CONACYT (278953-S0025-2016-1) project. Throughout the project we had the technical assistance of the Centro de Investigación en Ciencias de Información Geoespacial. For the technical support, we thank Sandra Medina and Gerardo Ávila, and specially we thank Gabriela Quiroz for the mapping making and visual design.

Author details

Juan Manuel Núñez*, Sandra Medina, Gerardo Ávila and Jorge Montejano
Centro de Investigación en Ciencias de Información Geoespacial AC (CentroGeo), Mexico City, Mexico

*Address all correspondence to: jnunez@centrogeo.edu.mx

References

[1] Seto KC, Satterthwaite D. Interactions between urbanization and global environmental change. Current Opinion in Environment Sustainability. 2010;**2**(3):127-128

[2] Núñez JM, Corona N, Ocampo P, Mohar A. Conectando el frente de agua marítimo de la zona costera norte de Yucatán con la zona metropolitana de Mérida. In: Iracheta A, Pedrotti C, Patricia R, editors. Suelo Urbano y Frentes de Agua: Debates y Propuestas en Iberoamérica. México: El Colegio Mexiquense, A.C.; 2017

[3] Caudillo C, Flores S. Tendencias espacio-temporales en la segregación. In: Tendencias territoriales determinantes del futuro de la Ciudad de México. México: Consejo Económico y Social de la Ciudad de México/Consejo Nacional de Ciencia y Tecnología/CentroGeo; 2016. pp. 153-175

[4] Mohar A. Tendencias territoriales determinantes del futuro de la Ciudad de México. Consejo Económico y Social de la Ciudad de México; 2016

[5] Núñez JM. Mapeo de la composición urbana, contraste entre dispersión y formas compactas en el sur de la Ciudad de México. In: Rothe HQ, editor. Ciudad Compacta: Del concepto a la práctica. Universidad Nacional Autónoma de México, Ciudad de México; 2015

[6] Batty M. The size, scale, and shape of cities. Science. 2008;**319**(5864):769-771

[7] Besussi E, Chin N, Batty M, Longley P. The structure and form of urban settlements. In: Remote Sensing of Urban and Suburban Areas. Berlin, Heidelberg, New York: Springer-Verlag; 2010. pp. 13-31

[8] Weng QH. Remote sensing of impervious surfaces in the urban areas: Requirements, methods, and trends. Remote Sensing of Environment. 2012;**117**:34-49

[9] Guindon B, Zhang Y, Dillabaugh C. Landsat urban mapping based on a combined spectral-spatial methodology. Remote Sensing of Environment. 2004;**92**(2):218-232

[10] Schneider A, Friedl MA, Potere D. A new map of global urban extent from MODIS satellite data. Environmental Research Letters. 2009;**4**(4):11

[11] Ridd MK. Exploring a V-I-S (vegetation-impervious surface-soil) model for urban ecosystem analysis through remote sensing: Comparative anatomy for cities. International Journal of Remote Sensing. 1995;**16**(12):2165-2185

[12] Deng CB, Wu CS. BCI: A biophysical composition index for remote sensing of urban environments. Remote Sensing of Environment. 2012;**127**:247-259

[13] Bhaskaran S, Paramananda S, Ramnarayan M. Per-pixel and object-oriented classification methods for mapping urban features using Ikonos satellite data. Applied Geography. 2010;**30**(4):650-665

[14] Zhou W, Troy A. An object-oriented approach for analysing and characterizing urban landscape at the parcel level. International Journal of Remote Sensing. 2008;**29**(11):3119-3135

[15] Zhang J, Foody GM. Fully-fuzzy supervised classification of sub-urban land cover from remotely sensed imagery: Statistical and artificial neural network approaches. International Journal of Remote Sensing. 2001;**22**(4):615-628

[16] Schneider A, Friedl MA, Potere D. Mapping global urban areas using

MODIS 500-m data: New methods and datasets based on 'urban ecoregions'. Remote Sensing of Environment. 2010;**114**(8):1733-1746

[17] Byun Y, Choi J, Han Y. An area-based image fusion scheme for the integration of SAR and optical satellite imagery. IEEE Journal of Selected Topics in Applied Earth Observations and Remote Sensing. 2013;**6**(5):2212-2220

[18] Puissant A, Zhang W, Skupinski G, editors. Urban morphology analysis by high and very high spatial resolution remote sensing. In: International Conference on Geographic Object-Based Image Analysis. 2012

[19] Duan YL, Shao XW, Shi Y, Miyazaki H, Iwao K, Shibasaki R. Unsupervised global urban area mapping via automatic labeling from ASTER and PALSAR satellite images. Remote Sensing. 2015;7(2):2171-2192

[20] Bartholome E, Belward AS. GLC2000: A new approach to global land cover mapping from earth observation data. International Journal of Remote Sensing. 2005;**26**(9):1959-1977

[21] Gao F, De Colstoun EB, Ma RH, Weng QH, Masek JG, Chen J, et al. Mapping impervious surface expansion using medium-resolution satellite image time series: A case study in the Yangtze River Delta, China. International Journal of Remote Sensing. 2012;**33**(24):7609-7628

[22] Esch T, Marconcini M, Marmanis D, Zeidler J, Elsayed S, Metz A, et al. Dimensioning urbanization—An advanced procedure for characterizing human settlement properties and patterns using spatial network analysis. Applied Geography. 2014;**55**:212-228

[23] Sandoval H, Núñez JM. Cuantificación de la composición biofísica de los ambientes urbanos de la ciudad de Mérida, Yucatán basada en el análisis de imágenes Landsat TM/ETM+/OLI (1986-2014). In: LCA C, LCB P, LCW Q, MET O, MIU C, MOG L, editors. Estudios Territoriales en México: Percepción Remota y Sistemas de Información Espacial. México: Universidad Autónoma de Ciudad Juárez; 2016

[24] Xian GZ. Remote Sensing Applications for the Urban Environment. Boca Raton, FL: CRC Press; 2015

[25] Maxwell AE, Warner TA, Fang F. Implementation of machine-learning classification in remote sensing: An applied review. International Journal of Remote Sensing. 2018;**39**(9):2784-2817

[26] Mas JF, Flores JJ. The application of artificial neural networks to the analysis of remotely sensed data. International Journal of Remote Sensing. 2008;**29**(3):617-663

[27] Dridi H, Bendib A, Kalla M. Analysis of urban sprawl phenomenon in Batna city (Algeria) by remote sensing technique. Analele Universităţii din Oradea, Seria Geografie. 2015;**2**:211-220

[28] Bhatta B. Analysis of Urban Growth and Sprawl from Remote Sensing Data. Berlin, Heidelberg, New York: Springer-Verlag; 2010. 170 p

[29] Ferreira FH, Messina J, Rigolini J, López-Calva L-F, Lugo MA, Vakis R. Economic Mobility and the Rise of the Latin American Middle Class. Washington, DC: The World Bank; 2012

[30] Richter R, Schlapfer D. Geo-atmospheric processing of airborne imaging spectrometry data. Part 2: Atmospheric/topographic correction. International Journal of Remote Sensing. 2002;**23**(13):2631-2649

[31] Jin HR, Stehman SV, Mountrakis G. Assessing the impact of training sample selection on accuracy of an urban classification: A case study in Denver, Colorado. International Journal of Remote Sensing. 2014;**35**(6):2067-2081

[32] Ghimire B, Rogan J, Galiano VR, Panday P, Neeti N. An evaluation of bagging, boosting, and random forests for land-cover classification in Cape Cod, Massachusetts, USA. GIScience & Remote Sensing. 2012;**49**(5):623-643

[33] Homer C, Huang CQ, Yang LM, Wylie B, Coan M. Development of a 2001 National Land-Cover Database for the United States. Photogrammetric Engineering and Remote Sensing. 2004;**70**(7):829-840

[34] Yu L, Liang L, Wang J, Zhao Y, Cheng Q, Hu L, et al. Meta-discoveries from a synthesis of satellite-based land-cover mapping research. International Journal of Remote Sensing. 2014;**35**(13):4573-4588

[35] Haykin S. Neural Networks: A Comprehensive Foundation. India: Prentice Hall PTR; 1994

[36] Atkinson PM, Tatnall A. Introduction neural networks in remote sensing. International Journal of Remote Sensing. 1997;**18**(4):699-709

[37] Canziani G, Ferrati R, Marinelli C, Dukatz F. Artificial neural networks and remote sensing in the analysis of the highly variable Pampean shallow lakes. Mathematical Biosciences and Engineering. 2008;**5**(4):691-711

[38] Basheer IA, Hajmeer M. Artificial neural networks: Fundamentals, computing, design, and application. Journal of Microbiological Methods. 2000;**43**(1):3-31

[39] Minsky M, Papert S. Perceptron Expanded Edition. Cambridge, MA: MIT Press; 1969

[40] Rosenblatt F. Principles of Neurodynamics: Perceptrons and the Theory of Brain Mechanisms. Washington, DC: Spartan; 1962

[41] Richards JA, Richards J. Remote Sensing Digital Image Analysis. Berlin, Heidelberg, New York: Springer-Verlag; 1999

[42] Fuller DO. Remote detection of invasive Melaleuca trees (Melaleuca quinquenervia) in South Florida with multispectral IKONOS imagery. International Journal of Remote Sensing. 2005;**26**(5):1057-1063

[43] Augusteijn MF, Folkert BA. Neural network classification and novelty detection. International Journal of Remote Sensing. 2002;**23**(14):2891-2902

[44] Liu WG, Wu EY. Comparison of non-linear mixture models: Sub-pixel classification. Remote Sensing of Environment. 2005;**94**(2):145-154

[45] Mertens KC, Verbeke LPC, Westra T, De Wulf RR. Sub-pixel mapping and sub-pixel sharpening using neural network predicted wavelet coefficients. Remote Sensing of Environment. 2004;**91**(2):225-236

[46] Lafont D, Guillemet B. Beam-filling effect correction with subpixel cloud fraction using a neural network. IEEE Transactions on Geoscience and Remote Sensing. 2005;**43**(5):1070-1077

[47] Mountrakis G, Im J, Ogole C. Support vector machines in remote sensing: A review. ISPRS Journal of Photogrammetry and Remote Sensing. 2011;**66**(3):247-259

[48] Vapnik V. Estimation of Dependences Based on Empirical Data. USA: Springer Science & Business Media; 2006

[49] Zhu GB, Blumberg DG. Classification using ASTER data and

SVM algorithms; the case study of Beer Sheva, Israel. Remote Sensing of Environment. 2002;**80**(2):233-240

[50] Pal M, Mather PM. Support vector machines for classification in remote sensing. International Journal of Remote Sensing. 2005;**26**(5):1007-1011

[51] Cortes C, Vapnik V. Support-vector networks. Machine Learning. 1995;**20**(3):273-297

[52] Boser BE, Guyon IM, Vapnik VN, editors. A training algorithm for optimal margin classifiers. In: Proceedings of the Fifth Annual Workshop on Computational Learning Theory. ACM; 1992

[53] Vapnik V. The Nature of Statistical Learning Theory. New York: Springer Science & Business Media; 2013

[54] Ustuner M, Sanli FB, Dixon B. Application of support vector machines for land use classification using high-resolution RapidEye images: A sensitivity analysis. European Journal of Remote Sensing. 2015;**48**:403-422

[55] Friedl MA, Brodley CE. Decision tree classification of land cover from remotely sensed data. Remote Sensing of Environment. 1997;**61**(3):399-409

[56] De Fries R, Hansen M, Townshend J, Sohlberg R. Global land cover classifications at 8 km spatial resolution: The use of training data derived from Landsat imagery in decision tree classifiers. International Journal of Remote Sensing. 1998;**19**(16):3141-3168

[57] Pal M. Random forest classifier for remote sensing classification. International Journal of Remote Sensing. 2005;**26**(1):217-222

[58] Sharma R, Ghosh A, Joshi PK. Decision tree approach for classification of remotely sensed satellite data using open source support. Journal of Earth System Science. 2013;**122**(5):1237-1247

[59] Burges CJC. A tutorial on support vector machines for pattern recognition. Data Mining and Knowledge Discovery. 1998;**2**(2):121-167

[60] Besag J. On the statistical analysis of dirty pictures. Journal of the Royal Statistical Society: Series B: Methodological. 1986;**48**(5-6):259-302

[61] Mesev V. Modified maximum likelihood classifications of urban land use: Spatial segmentation of prior probabilities. Geocarto International. 2001;**16**(4):41-48

[62] Majd MS, Simonetto E, Polidori L. Maximum likelihood classification of single high-resolution polarimetric SAR images in urban areas. Photogrammetrie, Fernerkundung, Geoinformation. 2012;(4):395-407

[63] Lucas I, Janssen F, van der Wel FJ. Accuracy assessment of satellite derived land cover data: A review. Photogrammetric Engineering and Remote Sensing. 1994;**60**(4):479-426

[64] Bharatkar PS, Patel R. Approach to accuracy assessment tor RS image classification techniques. International Journal of Scientific and Engineering Research. 2013;**4**(12):79-86

[65] Lu DS, Mausel P, Batistella M, Moran E. Comparison of land-cover classification methods in the Brazilian Amazon Basin. Photogrammetric Engineering and Remote Sensing. 2004;**70**(6):723-731

[66] Pal M, Mather PM. An assessment of the effectiveness of decision tree methods for land cover classification. Remote Sensing of Environment. 2003;**86**(4):554-565

[67] Lu D, Weng Q. A survey of image classification methods and techniques for improving classification performance. International Journal of Remote Sensing. 2007;**28**(5):823-870

[68] Huang Z, Lees BG. Combining non-parametric models for multisource predictive forest mapping. Photogrammetric Engineering and Remote Sensing. 2004;**70**(4):415-425

[69] Steele BM. Combining multiple classifiers: An application using spatial and remotely sensed information for land cover type mapping. Remote Sensing of Environment. 2000;**74**(3):545-556

[70] Liu W, Gopal S, Woodcock CE. Uncertainty and confidence in land cover classification using a hybrid classifier approach. Photogrammetric Engineering and Remote Sensing. 2004;**70**(8):963-971

[71] Lo CP, Choi J. A hybrid approach to urban land use/cover mapping using landsat 7 enhanced thematic mapper plus (ETM+) images. International Journal of Remote Sensing. 2004;**25**(14):2687-2700

[72] Kuemmerle T, Radeloff VC, Perzanowski K, Hostert P. Cross-border comparison of land cover and landscape pattern in Eastern Europe using a hybrid classification technique. Remote Sensing of Environment. 2006;**103**(4):449-464

[73] Carlson T. Preface—Applications of remote sensing to urban problems. Remote Sensing of Environment. 2003;**86**(3):273-274

Chapter 6

Water Management in Irrigation Systems by Using Satellite Information

Gema Marco Dos Santos, Ignacio Meléndez Pastor, Jose Navarro Pedreño and Ignacio Gómez Lucas

Abstract

Changes in agriculture are associated to the availability of resources and the economic and social demands. One of the more important transformations is to change rainfed into irrigated crops to increase the yield. In most of the cases, water resource and irrigation reservoirs are needed to maintain the yield. However, evaporation from ponds can be an important economic loss and an unsustainable strategy for water management, especially in arid and semiarid regions. Efficient methods for water storage should be established. In this study, a selected area located close to the city of Cartagena (Murcia) and the south of Alicante (Spain) has been studied, where there was an important transformation from rainfed to irrigated crops. Because of the high temperatures and insolation, the increment of the number of reservoirs detected by using remote sensing data and GIS tools may be inefficient for water management. The characterization of these reservoirs, to quantify the potential loss of water due to evaporation, has been done. The use of these tools for analysis could be interesting to find more efficient storage solutions (i.e., better spatial distribution of reservoirs, an increment of depth, and reduction of surface exposure) for improving the water storage and management.

Keywords: arid environments, evaporation, irrigated agriculture, spatial distribution, water storage

1. Introduction

Water management is one of the most important problems for future decades. Although there are areas of the planet where the water availability is naturally scarce due to the rain and temperature patterns, human pressure on this resource is accentuating the problem of scarcity. As reflected in the World Water Assessment Program published by UNESCO [1], there are three types of pressures or "drivers" on water systems: demographic, economic, and social. Population growth increases not only water consumption but also pollution, which is another way to decrease water availability. Furthermore, land occupation and urbanization affect the dynamics of the ecosystem due to soil sealing, and consequently, the hydrological cycle is altered (infiltration processes, aquifer recharge, etc.). Protecting ecosystems is highly important to maintain the goods and services they offer us, and it is so necessary for life. Economic growth has allowed the development of modern extraction and production techniques that aggravate water scarcity. Natural dynamics of water is affected; i.e., river flows are altered or the water table is reduced. The building of infrastructures that benefits the commerce of both products and services associated to water management has been increased. The change in the lifestyle of many countries is reflected in the amount of water consumed, principally in those in which access to drinking water is easy and immediate. In contrast, in developing countries where there is scarcity and water pollution, it is a great challenge. Therefore, there is a social inequality that must be resolved.

For example, in the case of arid and semiarid areas [2], in which the amount of available water is limited due to the shortage and irregularity of rainfall, the development of irrigated agriculture has caused an increase in pressure on water resources. This affects highly negatively the agriculture, which is one of the biggest users of water with respect to the total demand of water (almost 80%) [3]. In these areas, where water is a limited resource, population growth exerts a great negative pressure on it. Agriculture must be able to supply the population even though the availability of water is the limiting factor for food production [4]. To guarantee the continuous supply of water for irrigation, small ponds are built to store the water and manage it according to their needs [5]. These ponds are usually shallow constructions located near the crops that will supply. However, it seems that the management of these small reservoirs is based on the experience of the farmer and not on contrasted technical criteria [6]. Water is a limited and essential resource for life that has to be managed efficiently, equitably, and allow future generations to have access to it. Therefore, the current management model should be changed to make sustainable use of available water resources and develop strategies that promote savings and minimize losses in irrigation [7].

Evaporation is defined as a process by which liquid water turns to vapor state by heating it (energy breaks the bonds of the molecules) [8]. The main factors that influence evaporation are local climatic conditions such as air and water temperature, solar radiation, relative humidity, wind speed [9], and the geometry of the ponds, for example, evaporation is greater if the relationship between area and volume is large [10]. In areas with high insolation, the evaporation from the sheet of water represents a significant loss from the environmental and also economic point of view [11]. Different methods are being developed to avoid evaporation: there are chemical methods such as stearyl alcohol [12], floating modular systems that have different shapes and materials [13], floating photovoltaic panels [14], canvas, or suspended coverages [15]. Each method may be appropriate depending on the characteristics of the place where it will be installed (amount of water stored, area, costs, etc.) [16]. Therefore, it is necessary to study tools and develop management strategies that improve the efficiency of water consumption and obtain the potential evaporation from the ponds and reservoirs.

The use of Geographic Information Systems (GIS) in the study of water resources allows us to know the dynamics of them, and therefore, models with different scenarios of water availability or demands can be developed [17]. With these models, different projections can be made in order to develop management scenarios more suited to the state of resources. This technology, GIS, is very suitable due to the amount of information that can be incorporated into the models, and the possibility of viewing the information in the form of maps [18]. In developing countries, this tool can help the management of its resources with a relative low cost and the large number of free images obtained over many years from remote sensing. Moreover, in those countries in which it is not possible to collect data in situ because of the cost, time, or access due to legal causes or because of war conflicts.

GIS combining remote sensing help to solve many problems related to resources management.

Remote sensing is being a very useful method to delimit and study water bodies, especially due to the difficulty of obtaining continuous information. Due to the contrast between the reflectance of the sheet of water and that of the earth surface [10], it is possible, through satellite images, to study and monitor the water storage [19], to observe the changes in the surface of water bodies over time, study the evolution of the irrigation reservoirs of an area [20], estimate its evaporation (important in arid and semiarid zones) and volume. The water absorbs the energy in wavelengths of the near and medium infrared; therefore, the reflected energy of these is low and the water bodies appear in dark color in both the multispectral images and the grayscale images [21]. Moreover, satellite images facilitate the composition of RGB or false color images where water sheets can be detected and analyzed.

Facing of future scenarios of climate change [22], in which the availability and quality of water can be seriously affected [23], it is necessary to improve the use of water resources through the incorporation of new techniques and the modernization of infrastructures. This includes the application of regulations [24] that support integrated management techniques that guarantee a better resource quality and also promote citizen participation [25].

In this work, the combined use of remote sensing data and GIS tools, demonstrated with the example, the possibilities of managing and controlling water infrastructures and the evaporation of water in agriculture, is one of the major consumers of water.

2. A study case: Campo de Cartagena, southeast of the Iberian Peninsula

2.1 Study area

The study area is located beside the Mar Menor in Murcia and south of Alicante (**Figure 1**), Spain. This basin is a sedimentary plain formed by conglomerates, marls, sandstones, and clays [26] with approximately 152,000 ha. The Mar Menor is the biggest coastal lagoon of Spain that is included in the RAMSAR convention. It is in serious danger of pollution as a result of nitrogen and phosphate contributions from agriculture that cause the loss of its water's quality, the decrease of the diversity and elimination of autochthonous species, and induce the proliferation of algae blooms. The two factors that most affect this wetland are tourism (population growth) and agriculture; both generate polluting inputs that reach the Mar Menor through the different watercourses and infiltration processes. The climate is Bsh according to the Köppen classification, with low rainfalls (around 300 mm per year) of torrential type especially during the autumn. The average annual temperature is about 18°C, with hot summer (about 32–35°C in August) and mild winters (the temperature usually does not drop below 5°C) [27]. Precisely, the weather is one of the main reasons why so many tourists (both Spanish and foreign) come every year to visit the Region of Murcia (more than 1 million people in 2015–2016) [28], especially near the coast.

Different improvements, mainly since the second half of the twentieth century in the region of Murcia and Alicante province, have favored the growth of population, principally located in coastal areas. This increase may be due to the improvement of communication channels (roads) and greater availability of water resources, which has allowed the development of agriculture. Agriculture is very important in the Region of Murcia, because of the good climate and a fertile soil

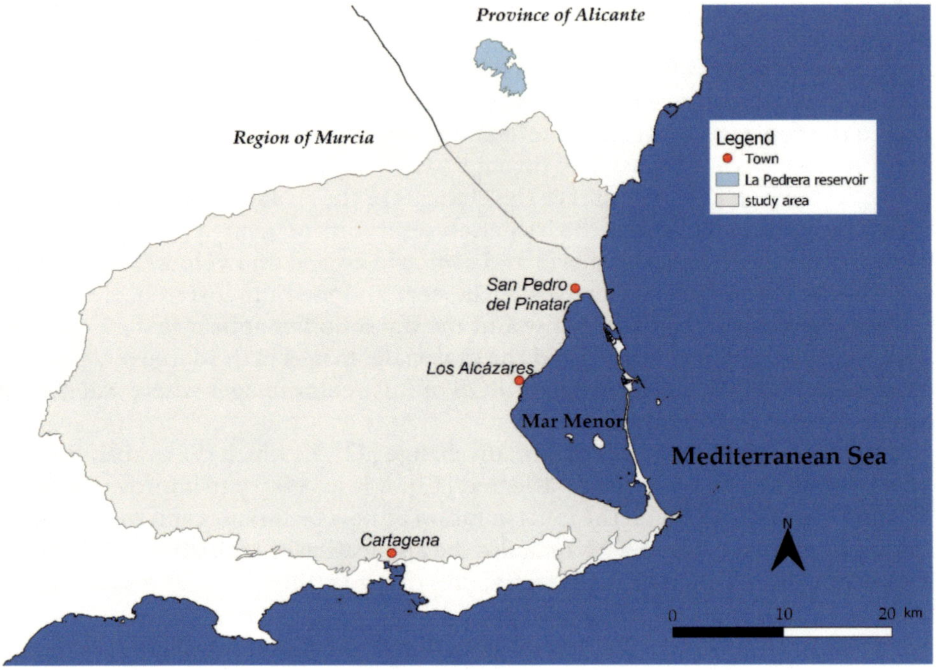

Figure 1.
Location of study area (Campo de Cartagena, Mar Menor watershed) in the region of Murcia and the south of the province of Alicante.

in many river basins that allows suitable growth of crops, but the lack of water has limited the production. Therefore, the change of rainfed crops to irrigated crops was benefited by the capital investment (the Tajo-Segura water transfer in 1979, the exploitation of the aquifers and the obtaining of desalinated water), which increases the availability of water; the productivity of the crops has improved in spite of the severe shortage suffered by the area. La Pedrera reservoir (built in 1985), located in the province of Alicante, is responsible for regulating the water, which comes from the Tajo-Segura transfer canal (agricultural and urban supply). This reservoir maintains adequate water availability despite the severe scarcity suffered in the area [29]. In fact, Murcia exports between 20 and 30% of fruits and vegetables in Spain, especially to the European Union [30]. Even with the external contributions of water, it is not enough to supply the water needs of the area that often suffers serious droughts that cause cuts back not only for agriculture but also for urban supply. In addition, during the summer, the demand for water for agriculture is higher because of the large water deficit and high temperatures. This situation also coincides with the period of greatest urban demand in the area due to tourism [31], especially in some areas closer to the coast. For example, it is estimated that on the Costa Cálida, there were almost 4 million visitors in 2016 [28].

2.2 Methodology

The data were obtained from the National Geographic Institute (IGN). We used the geodesic reference system ETRS89 and UTM projection zone 30 [32].

The Mar Menor watershed was delimited with the Digital Terrain Model (MDT25 CC-BY 4.0 scne.es) and the GRASS software using the flow lines that run along the maximum slope. The basin covers 151,641 ha and is located mainly in the Region of Murcia and a part in the province of Alicante. All the reservoirs of the basin were digitized, one by one by, using high-quality orthophotos from the Plan

Figure 2.
Old mills and the new irrigation systems in the Campo de Cartagena.

Nacional de Ortofotografía Aérea (PNOA) (FotoPNOA 2004–2016 CC-BY 4.0 scne.es, pixel size of 25 cm). The same process was followed with old photographs taken from a photogrammetric flight along the period from 1973 to 1986 (Fotol 1973–1986 CC-BY 4.0 scne.es, scale 1: 18,000, pixel size between 27 and 45 cm). The digitization process was done with QGIS v.3.2. The ponds and reservoirs were marked with points to locate them, and then, they were digitalized to determine their surface taking into account the limits of the structure, when they were at maximum capacity. A field trip was also done to compare the results obtained from the images with the disposition in fact, checking close to a hundred elements (old mills, ponds, and small reservoirs) (**Figure 2**).

A heat map was developed from the density of points that identify the location of each irrigation ponds/reservoirs to better understand their distribution. Point interpolation aids to visualize in a map the concentration of these in a continuous surface. Three parameters are used to create a heat map: the cell size, the bandwidth, and the type of calculation used in the interpolation. The cell size will determine the degree of detail on the surface. The larger the cell size, the less continuous the color gradient that represents the concentration of points will be. The bandwidth (or search radius) is the area around each point that the GIS will take into account for density calculation. The type of calculation used in the most common interpolation is inverse distance weighting (IDW), which assigns more importance to the functions that are closer than to those that are furthest away [33]. In this case, we used a search radius of 5 km and 15 pixels of cell size.

To estimate annual evaporation losses in the study area, we have used as reference the evaporation values published in the article "Regional assessment of evaporation from agricultural irrigation reservoirs in a semiarid climate" by Martínez Alvarez et al. [34]. They use measures done in the 2003 for the entire Segura River basin (located in the southeast of Spain, including the study area). In this article, authors estimate the evaporation losses using daily, monthly, and annual data on temperature, precipitation, relative humidity, wind speed, wind direction, and solar radiation of 74 agro-meteorological stations for the period 2000–2006. In addition, some of them have class-A pan evaporometer in which evaporation was calculated by a sensor that determinate the difference in water level. The class-A pan evaporometer standardized by the US National Weather Service is a 120.7-cm diameter and 15-cm-deep cylinder made of galvanized iron. It is elevated about 15 cm from the ground by a wooden platform. It must be located where the air circulates freely so that it does not affect the measurements [35]. They use 14,145 irrigation reservoirs for the entire Segura basin, which occupied 4901 ha. They obtain as a result the annual evaporation loss in the Segura basin taking into account the maximum surface area, which was 68.8 hm^3. Based upon this value and considering the surface, the evaporation value of water used as a reference is 0.014 hm^3 year/ha. This helps us to estimate the evaporation loss estimation in our study area.

3. Results and discussion

Figure 3 presents the digitized points that indicate the location of the irrigation ponds for both periods. There is a clear increase in the number of points currently with respect to the previous period.

In the image a (**Figure 3**), the points do not appear distributed following any regular pattern; they are dispersed throughout the basin but especially near the coast and the urban cores, some of them forming small groups. In the top of the basin (NW), in the foothills of the Sierra de Carrascoy and El Valle, there are no irrigation ponds because at that time, mechanization and cultivation techniques did not allow working the land in areas with steep slopes. In the image b, there is a greater increase of irrigation ponds and small reservoirs. Grouping of points can be observed mainly in the center of the basin, which is quite flat, and in the top near La Pedrera reservoir. There is also a tendency for a high density of points near the coast as in the first image. In this case, due to the modern techniques, the irrigated crops occupy the foothills of the mountains.

In order to understand and visualize better the irrigation ponds distribution patterns in the area and compare them between two periods, a heat map (**Figure 4**) was created from the density of points. These maps confirm in a very clear way the changes produced in the area.

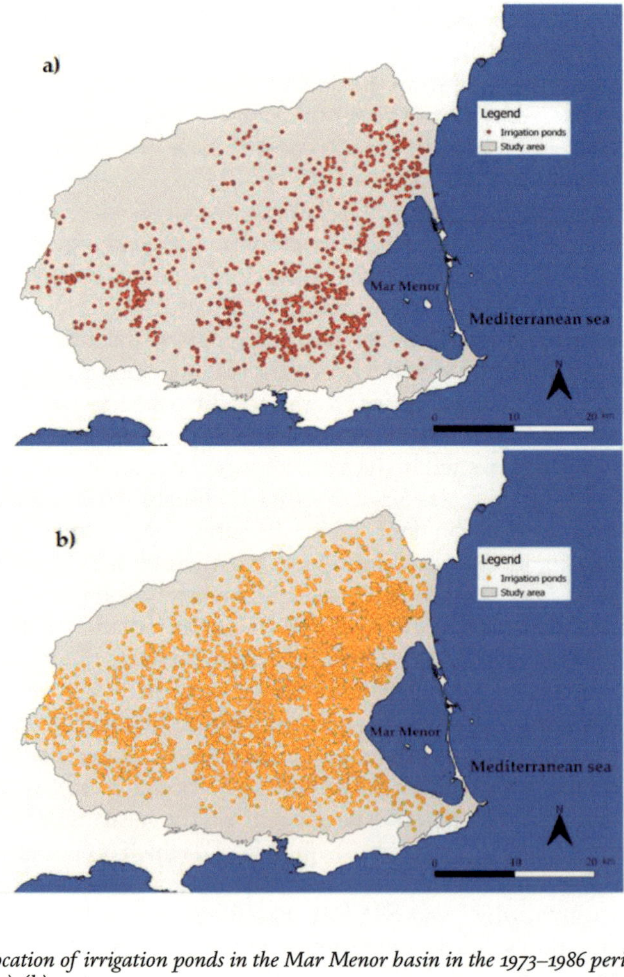

Figure 3.
Points marking the location of irrigation ponds in the Mar Menor basin in the 1973–1986 period (a) and nowadays (2016–2017) (b).

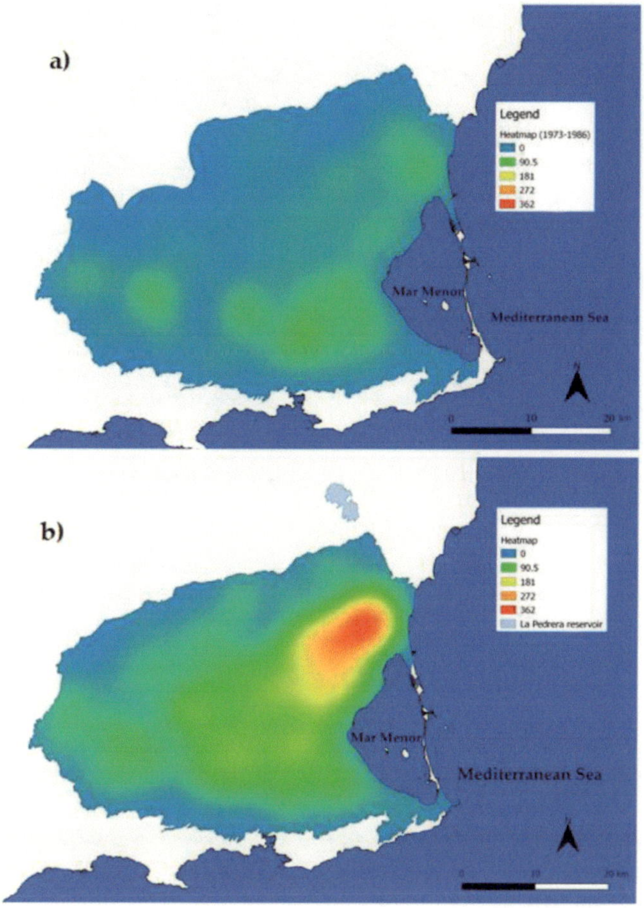

Figure 4.
Heat maps from dot density of the identified ponds with old photographs (a) and those with current ones (b).

In the first image (**Figure 4a**), high density of points (in green) is observed in the lower part of the basin and following the coastline. This location may be associated with the extraction of water from the subsurface aquifers, following the pattern of the traditional systems such as windmills. (A pond was situated near the mill so that the water could fall into it.) Moreover, extraction that is more efficient with pumps made possible to obtain water from the aquifers at a larger depth coming to cause an overexploitation of aquifers.

The arrival of water from the Tajo-Segura transfer in 1979 increased the availability of water and relieved the pressure on groundwater [36]. This situation benefited production and the expansion of intensive agriculture (with the corresponding construction of small reservoirs to store and supply water).

In the second image (**Figure 4b**), there is a generalized increment near the coast and a great increase in the upper part of the basin (NE). This difference could be explained by the construction in 1985 of the La Pedrera reservoir. It has 1272 ha and can store 246 hm^3. This reservoir receives water from the Tajo-Segura transfer and distributed to the Campo de Cartagena by a great canal and others supplied conductions.

La Pedrera reservoir is also used for urban water supply through the Taibilla canal. Therefore, it is easier to supply the crop fields closest to the reservoir. Consequently, it has favored a greater development of greenhouses (**Figure 5**). They are grouped near the towns of San Pedro del Pinatar (Murcia) and Pilar de la Horadada in the south of Alicante.

Figure 5.
Group of greenhouses near San Pedro del Pinatar in the study area (source: derived from FotoPNOA 2004–2016 CC-BY 4.0 scne.es).

	Number of ponds	Total area occupied (m^2)	Mean area per pond (m^2)	Estimated evaporation (hm^3/year)
1973–1986 period	971	886,349	318	1.24
2016–2017 period	3846	12,013,189	1631	16.82

Table 1.
Values obtained from the irrigation pond digitalization and estimated values of evaporation in the study area for both periods.

After the analysis of the data and the digitalization of the irrigation ponds from the images in both periods, **Table 1** shows a summary of them. A total of 971 reservoirs were digitized from the data of the period 1973–1986. The sheet of water, according to the sum of the surface of all them, accounted to 88.63 ha. The average surface area of the reservoir/pond was 318 m^2.

In the second period (2016–2017), 3846 irrigation ponds were digitized from PNOA images. The total water surfaces were 1201.32 ha. The average surface area of the reservoirs increased to 1631 m^2.

These values indicate that the number of reservoirs in this area has almost incremented four times. For the average surface of ponds, the size at present is five times higher than before, however not necessarily deeper than the oldest. Therefore, the total area occupied by the reservoirs has increased fourteen times and the size of the surface of the reservoirs only five times for the last four decades.

To estimate the possible evaporation losses from the sheet of water of the small reservoirs/ponds, we took as reference the value given for the area of 0.014 hm^3 year/ha [34]. Considering all the reservoirs to their maximum capacity, the values estimated for each period were as follows:

- For the 971 ponds in the first period (determined from the images obtained between 1973 and 1986), the annual losses would be close to 1.24 hm^3/year,
- and for the second one (mages of PNOA 2016–2017), it would be close to 16.82 hm^3/year.

This means that there is a difference of approximately 15.62 hm^3/year, parallel to the increment of the surface exposure of reservoirs and ponds. This amount of water that can be lost is equal to that needed for the supply of a city of 300,000 inhabitants for a year considering the average water consumption in Spain for inhabitants [37].

Water scarcity in this area has always been a main concern for agricultural production. However, with the transfer from other river basins (i.e., Tajo river), water availability has been increased and along with the population growth and agricultural yield. This was reflected in the construction of reservoirs/ponds in the last years, which has been increased. With this increment, the potential evaporation of water from reservoirs and ponds has been dramatically increased by the way.

According to a report managed by the Ministry of Agriculture and Water of the Region of Murcia with data from the Space Agency of Meteorology (AEMET), and with the collaboration of different universities and institutions, the evolution of rainfall does not follow a clear trend, which is a normal situation in that area with such irregularities. For temperatures, a slight tendency to increase is observed. In fact, according to this report from 1971 to 2009, the average annual temperature of the entire Region of Murcia increased from 15.5 to 17°C [38]. Therefore, the evaporation loss could be aggravated considering the scenarios based on the climate change and the increase in temperature. In this sense, evaporation can be over the values estimated in this work.

In this line, it is important to study and develop measures to avoid water evaporation and improve the efficiency of the irrigation system. For this reason, it is convenient to study a better spatial distribution of reservoirs and reduce the number of them. Moreover, an increment of depth in their construction can facilitate to store the same amount of water with less surface exposure to evaporation. Finally, the use of some techniques to cover the ponds can reduce the water surface exposure.

4. Conclusions

Remote sensing data are very useful to study and analyze the amount of water stored and the management of irrigation systems. The use of these technologies, both GIS and remote sensing, can help in the management of decision-making about water resources.

The example given shows that the amount of water that could evaporate represents a significant loss. In this case, the amount of water that could evaporate is almost 14 times higher now. This matches with the increase in the total surface occupied by the irrigation ponds. With only a four-time increment in the number of reservoirs, the amount of water that could evaporate increases by 350%. Although it is an estimation, it is clear that water losses due to evaporation represent a high cost, especially in areas where this resource is scarce.

Despite the water limitations of the area, in the Mar Menor basin, there are many agricultural fields that generate tons of fruits and vegetables that provide a great social and economic benefit to region. Even with the different sources of water, there is still a water deficit that generates (especially during droughts periods) economic, social, and environmental instability.

In addition, with the possible effects of climate change that indicates less precipitation and higher temperatures, it is expected that the amount of water resources available can be seriously affected especially in arid and semiarid areas such as Murcia and Alicante, which already suffer the effects of scarcity. Efforts should be done applying techniques to reduce the evaporation. Therefore, saving the resource to avoid losses as much as possible and be able to supply a growing population is a priority.

Acknowledgements

Thanks to the Instituto Nacional de Geografia (IGN) for the availability of social, economic, and environmental information in open source that facilitates the research in Spain.

Conflict of interest

Authors expressed that there is no conflict of interest.

Author details

Gema Marco Dos Santos, Ignacio Meléndez Pastor, Jose Navarro Pedreño* and Ignacio Gómez Lucas
Department of Agrochemistry and Environment, University Miguel Hernández of Elche, Alicante, Spain

*Address all correspondence to: jonavar@umh.es

References

[1] Nations U. Water in a Changing World [Internet]. Vol. 11, World Water. 2009. 349 p. Available from: http://www.esajournals.org/doi/abs/10.1890/1051-0761(2001)011[1027:WIACW]2.0.CO;2

[2] Martínez Fernández J, Selma MAE. The dynamics of water scarcity on irrigated landscapes: Mazarrón and Aguilas in South-Eastern Spain. System Dynamics Review. 2004;**20**(2):117-137

[3] European Environment Agency. El agua en la agricultura. 2012. Available from: https://www.eea.europa.eu/es/articles/el-agua-en-la-agricultura

[4] El-Beltagy AMM. Impact of climate change on arid lands agriculture. Agriculture & Food Security. 2012;**1**:3

[5] Juan M, Casas J, Bonachela S, Fuentes-Rodríguez F, Gallego I, Elorrieta MA. Construction characteristics and management practices of in-farm irrigation ponds in intensive agricultural systems—Agronomic and environmental implications. Irrigation and Drainage. 2012;**61**(5):657-665

[6] Cazorla MJ. Gestión ecosistémica de las balsas de riego del litoral mediterráneo andaluz [thesis]. Almería University; 2012

[7] Hamdy A, Abu-Zeid M, Lacirignola C. Water crisis in the mediterranean: Agricultural water demand management. Water International. 1995;**20**(4):176-187

[8] USGS. The Water Cycle: Evaporation [Internet]. 2016. Available from: https://water.usgs.gov/edu/watercycleevaporation.html

[9] Benzaghta MA, Mohamad TA. Evaporation from reservoir and reduction methods: An overview and assessment study. Domascus, Syria Medinah, Kingdom Saudi Arab: Int Eng Conv; 2009. 9 p

[10] Ibarra D, Salvador M, Conesa C. Estimación de evaporación en balsas de riego mediante el empleo de técnicas de teledetección. Estudio aplicado a la vertiente litoral sur de la Región de Murcia. Vol. 82014

[11] Lopez Moreno JI. Estimación de pérdidas de agua por evaporación en embalses del Pirineo. Cuadernos de Investigación Geográfica. 2008;**34**:61-81

[12] Gugliotti M, Baptista MS, Politi MJ. Reduction of evaporation of natural water samples by monomolecular films. Journal of the Brazilian Chemical Society. 2005;**16**(6 A):1186-1190

[13] Segal L, Burstein L. Retardation of water evaporation by a protective float. Water Resources Management. 2010;**24**(1):129-137

[14] KYOCERA TCL Solar Begins Operation of Japan's Largest 13.7 MW Floating Solar Power Plant [Internet]. 2018. Available from: http://www.kyocerasolar.eu/index/news/news_details.L3NvbGFyX2VsZWN0cml jX3N5c3RlbXMvbmV3cy8yMDE4 L0tZT0NFUkFfVENMX1Nvb GFyX2JlZ2luc19vcGVyYXRpb25fb2Z fSmFwYW5fc19sYXJnZXN0Xz EzXzdNV19GbG9hdGluZ19Tb2xhc l9Qb3dlcl9QbGFudA~~.html

[15] Yao X, Zhang H, Lemckert C, Brook A. Evaporation Reduction by Suspended and Floating Covers: Overview, Modelling and Efficiency. Urban Water Security Research Alliance Technical Report No. 28; 2010

[16] Craig I, Green A, Scobie M, Schmidt E. Controlling Evaporation Loss from

Water Storages. Natl Cent Eng Agric. 1000580/1; 2005. p. 207

[17] Kawsar R. Water Resource Managment and Remote Sensing, A Prospective Issue that Requires Considerable Attention [Internet]. 2015. Available from: http://geoawesomeness.com/water-resource-managment-and-remote-sensing-a-prospective-issue-that-requires-considerable-attention/

[18] Tsihrintzis VA, Hamid R, Fuentes HR. Use of geographic information systems (GIS) in water resources: A review. Water Resources Management. 1996;**10**(4):251-277

[19] Pipitone C, Maltese A, Dardanelli G, Lo Brutto M, La Loggia G. Monitoring water surface and level of a reservoir using different remote sensing approaches and comparison with dam displacements evaluated via GNSS. Remote Sensing. 2018;**10**(1):1-24

[20] Lin YCW, Hsiao L, Cheng K. A Multi-Decadal Change Analysis for Irrigation Ponds in Taoyuan, Taiwan Using Multi-source Data2018. pp. 1-16

[21] Kite G, Pietroniro A. Remote sensing of surface water. In: Schultz GA, Engman ET, editors. Remote Sensing in Hydrology and Water Management. Berlin, Heidelberg: Springer; 2000. pp. 217-238

[22] Porter JR, Xie L, Challinor AJ, Cochrane K, Howden SM, Iqbal MM, et al. Food security and food production systems. In: Field CB, Barros VR, Dokken DJ, Mach KJ, Mastrandrea MD, Bilir TE, Chatterjee M, Ebi KL, Estrada YO, Genova RC, Girma B, Kissel ES, Levy AN, MacCracken S, Mastrandrea PR, White LL, editors. Climate Change 2014: Impacts, Adaptation, and Vulnerability Part A: Global and Sectoral Aspects Contribution of Working Group II to the Fifth Assessment Report of the Intergovernmental Panel on Climate Change. Cambridge, United Kingdom and New York: Cambridge University Press; 2014. pp. 485-533

[23] Hudson NW. Water Conservation. In: FAO Land and Water Development Division, editor. Soil and Water Conservation in Semi-Arid Areas. Bedford, United Kingdom; 1988. p. 185

[24] Vargas-Amelin EPP. The challenge of climate change in Spain: Water resources, agriculture and land. Journal of Hydrology. 2014;**518**:243-249

[25] FAO & WWC. Towards a Water Critical Perspectives for Policy-makers. FAO. 2015. p. 62

[26] Pérez Ruzafa A, Marcos C, Pérez Ruzafa IM. 30 años de estudios en la laguna costera del Mar Menor: de la descripción del ecosistema a la comprensión de los procesos y la solución de los problemas ambientales. In: Instituto Euromediterráneo del agua, editor. El Mar Menor: Estado del conocimiento actual. 1st ed. 2009. pp. 18-40

[27] Blázquez MP, Pelegrín GB, Díaz MF. Ficha informativa de los humedales Ramsar. Mar Menor; 2006

[28] Mompeán PA, Vegas Juez AM. Turismo en la región de murcia 2016. Instituto de Turismo de la Región de Murcia: ITREM; 2016

[29] Confederación Hidrográfica del Segura. Embalse de La Pedrera [Internet]. 2018. Available from: https://www.chsegura.es/chs/cuenca/infraestructuras/embalses/embalsedelaPedrera/index.html

[30] FEDEX. Exportación/importación españolas de frutas y hortalizas [Internet]. 2018. Available from: http://www.fepex.es/datos-del-sector/produccion-frutas-hortalizas

[31] Candela L, Domingo F, Berbel JAJJ. An overview of the main water

conflicts in Spain: Proposals for problem-solving. In: El Moujabber M, Ouessar M, Laureano P, Rodríguez R, editors. Water Culture and Water Conflict in the Mediterranean Area. Options Méditerranéennes: Série A. Séminaires Méditerranéens; no. 83. 2008. pp. 197-203

[32] Real Decreto 1071/2007, de 27 de julio, por el que se regula el sistema geodésico de referencia oficial en España

[33] Dempsey C. Heat Maps in GIS [Internet]. GIS Lounge. 2012. Available from: https://www.gislounge.com/heat-maps-in-gis/

[34] Martínez Alvarez V, González-Real MM, Baille A, Maestre Valero JF, Gallego Elvira B. Regional assessment of evaporation from agricultural irrigation reservoirs in a semiarid climate. Agricultural Water Management. 2008;**95**(9):1056-1066

[35] Toribio MIS. In: Sociedad Española de Geomorfología, editor. Métodos para el estudio de la evaporación y evapotranspiración. 1st ed. Logroño: Geomorga Ediciones; 1992

[36] Gil Meseguer E. Los paisajes agrarios de la región de Murcia. Papeles de Geografía. 2006;**43**:19-30

[37] Instituto Nacional de Estadística. España en cifras 2017. Madrid; 2018

[38] Victoria Jumilla F. Cambio Climático en la Región de Murcia. Murcia: Consejería de Agricultura y Agua; 2010

Section 5

Spatial Data for Natural Features Monitoring

Chapter 7

Validation of Satellite (TMPA and IMERG) Rainfall Products with the IMD Gridded Data Sets over Monsoon Core Region of India

Tumuluru Venkata Lakshmi Kumar, Humberto Alves Barbosa, Manoj Kumar Thakur and Franklin Paredes-Trejo

Abstract

This work presents the validation of satellite (TMPA and IMERG) rainfall products against the India Meteorological Department (IMD) gridded data sets (0.25° × 0.25°) of dense network of rain gauges distributed over the monsoon core region of India. The validation uses the data sets covering the 20 years (1998–2017) and detects the time series bias; inter annual variations and Intra Seasonal Oscillations (ISO). The bias in the two data sets is found to be very less over the core region compared to whole India. The correlation between daily rainfall IMD and satellite is found to be +0.88 which is of 99% confidence level. The dominant periodicities in the rainfall patterns of IMD and satellite are Madden Julie Oscillation (30–60 days) and local oscillations (less than 20 days) are conspicuous and the normalized power varies from year to year. During the El Niño and La Niña years, the normalized power of rainfall pattern is low and high in satellite data sets which infer the suppressed and strongest activity of MJO over Indian Ocean that modulates the rainfall pattern over India.

Keywords: TMPA, IMERG, IMD gridded rainfall, ISO, validation

1. Introduction

During the recent decades, a number of remotely sensed rainfall products have been developed from satellites and are being used widely for different applications such as weather forecasting, hydrology and water resource management. The satellite rainfall data is also useful in assessing the large scale droughts [1] and to monitor the extreme weather events which is increasing due to climate change [2, 3]. The estimation of rainfall from satellites mainly depends upon the relationship between rain rate and the cloud top temperature, observed from the infrared sensors and the influence of rain drops on microwave radiation. The IR based techniques apply the empirical methods to obtain the rainfall from the cloud top brightness temperature, assuming the rainfall originates from the convective clouds [4, 5]. On the other way, microwave techniques directly sense the radiation emission and scattering occurring due to presence of hydrometeors and provide the rainfall estimates. The lacuna in the two methods are IR technique fails in the case of warm clouds and MW measurements are with less frequency compared to IR passes [6]. However, the combination of the two aforementioned techniques provide the best estimates of rainfall, which was employed in the retrieval algorithms of TRMM Multi-satellite Precipitation Analysis (TMPA) and Climate Prediction Centre (CPC) morphing technique (CMORPH). The TRMM was launched as the joint effort of National Aeronautics and Space Agency (NASA), USA and Japan Aerospace Exploration Agency (JAXA), Japan to monitor and assess the tropical rainfall and associated latent heating [7]. The main TRMM sensors are a TRMM Microwave Imager (TMI)], precipitation radar (PR) and a visible and infrared scanner (VIRS). These are combined variously with other IR and gauge-based products [8]. The unique advantage with TRMM that it passes through all hours, thus the diurnal variation of rainfall can be studied. The 3-hourly data records of TRMM rainfall with 0.25 grid resolution are made available from the year 1998 to till 2013 and thereafter, the Global Precipitation Measurement (GPM), the improved version of TRMM which allowed to better link the data sets were been launched [9]. Arrival of GPM brought revolutionary changes in the studying the rainfall characteristics of the convective systems, storms etc. Having with the high spatiotemporal resolution of 0.10 degree and half hourly precipitation products, the GPM rainfall estimates are proven to be more reliable to study the characteristics of tropical cyclones and other rainfall induced hazardous events [10].

Nevertheless, the satellite rainfall products cannot accurately estimate the rainfall. They are much helpful over large areas when there is a limited coverage of rain gauges which of their average may not represent the whole areal rainfall picture in terms of its variability and magnitude. Over small scale areas, the satellites show inability to capture the localized variations. Hence, it ought to validate the satellite rainfall with the ground based measurements of any area before is being used. Several studies focuses globally to validate the TMPA rainfall products not only with the ground based measurements but also with the existing reanalysis products and reported various issues among the comparison results. A study over Caspian Sea Region of Iran by Duan et al. [11] showed that the TRMM 3B42 version rainfall data replicates the monthly and annual variations as gauge data. They also reported that TRMM underestimates the heavy rainfall events over Iran. Nair et al. [12] showed that TRMM 3B42 V6 rainfall products were unable to estimate the accurate amount of orographic rainfall (over Western Ghats of India) and it was underestimated over the rain shadow regions. But the satellite is able to capture the rainfall gradient from west to east of Western Ghats region. It is also reported that the TRMM algorithm was unable to pick the average high and low daily rainfall over India [13]. Uma et al. [14] compared the TRMM 3B42 V6 rainfall data with the India Meteorological Department (IMD) gridded data developed by [15] from 1998 to 2007 and found that at 5° × 5° scales and beyond, both products are compatible to each other. The temporal scale assessment of them showed the pentads of TRMM rainfall (TMPA) are in good agreement with the IMD gridded rainfall data over India. Cao et al. [16] performed the evaluation of 3B43 TRMM data with the ground measurements of rainfall over Yangtze River Delta of China during the period 1998–2016. The results of their study infer that 3B43 data overestimates the actual precipitation but maintained the consistency in terms of correlation and bias. The GPM which is the successor of TRMM offering half hourly rainfall products were also been validated over different regional scales. The comparison of GPM rainfall product IMERG (Integrated Multi satellite Retrievals for GPM) with TRMM 3B42 V7 products over southern Tibetan Plateau indicates the GPM outperforms the TRMM in all spatial scales [17]. GPM shows better detecting capacity than TRMM during light rainy days. Preliminary results of assessment of GPM over India show the difficulties in detecting the rainfall events over south east peninsular and north eastern parts of India [18]. Though there are numerous studies on the validation and comparison of TRMM, GPM rainfall amounts, studies on how satellite rainfall is able to capture the dominant periodicities and interannual variability caused by the global teleconnections such as El Niño and La Niña are sparse. The present work has been mainly focused on the consistency and ability of satellite (TRMM and GPM) in picking the interannual variations and dominant periodicities of Indian south west monsoon. The study on the periodical features of rainfall derived from the satellites are very important in understanding the large scale circulation patterns and have implications on the soil erosion where ground data is not available [19]. Henceforth, we focus on the inter-annual variations and prevailing seasonal oscillations of Indian summer monsoon rainfall obtained from the TRMM/GPM and IMD gridded data sets for the period 1998–2017.

2. Data and methods

The daily rainfall data from TRMM 3B42 V7 for the period 1998–2013 has been used in the present study along with the IMERG data from 2014 to 2017. The TRMM data is available at 0.25° × 0.25° degree spatial resolution published by the Goddard Space Flight Centre Distributed Active Archive Centre (GSFC DAAC). IMERG half hourly data is available with 0.1° × 0.1° spatial resolutions and the same is considered from 2014 to 2017. Both the data sets have been averaged to 0.25° × 0.25° degree spatial resolution and used in the present study for the period from 1998 to 2017. The rainfall data has been extracted to the monsoon core region of India [20] on daily basis and converted to monthly scales by cumulating the daily rainfall for the months of south west monsoon (June, July, August and September) for the period 1998–2017. Before the onset of southwest monsoon, the core monsoon region is characterized by a heat low and thereafter, establishment of a tropical convergence zone takes place in this region [21]. The boundary from the heat low areas over the northwestern part of India to the moist convective parts of the eastern part of the monsoon zone also exhibit significant variation from year to year. The region has different characteristics during a poor and good monsoon years. The monsoon core region of India is shown in the **Figure 1** and the rainfall data pertaining to this region is used for the present analysis.

IMD gridded data with 0.25° × 0.25° spatial resolution spread over Indian land mass has been used as the ground validation in the present study. These data sets were developed by Pai et al. [22] based on the rainfall record from 6955 rain gauges located in India. Out of 6955 rain gauges, 547 are from IMD; 494 from hydro-meteorology; 74 from agromet and remaining from the stations maintained by the state government India. The average number of stations used to obtain 1 day rainfall is about 3100. This version of data is known as IMD4 whereas the earlier versions of this data, IMD3, IMD3 and IMD1 are developed for the periods 1971–2005, 1901–2004 and 1951–2007 0.5° × 0.5° and 1° × 1° spatial resolutions over Indian landmass and 6076, 1380 and 2140 number of rain gauges were used. The dense network of rain gauge locations are depicted in the **Figure 1** [22]. After collecting the station rainfall data, Shepard interpolation technique has been applied to prepare the gridded data sets and the detailed procedure of processing the data can be found from Pai et al. [22].

The rainfall data from the TRMM and GPM have been downloaded from the website http://mirador.gsfc.nasa.gov. The data sets are available in nc4 data format and are processed using the MATLAB as suggested in the http://disc.gsfc.nasa.gov. **Table 1** gives the complete information on the details of rainfall data sets used from TMPA and IMERG including their algorithms.

In the present work, we combine the two satellites rainfall data of TRMM and IMERG covering the period from 1998 to 2017. TMPA 3B42 V7 data sets for the period 1998–2013 and IMERG rainfall products for the period 2014–2017 were used. Each grid value of rainfall over the monsoon core region of India is averaged with the number of grid points and is allotted as the rainfall of the region for the day. Likewise, we computed the daily rainfall over the study region for the southwest monsoon months of June to September for the year from 1998 to 2017. The flow chart given below illustrates the data retrieval methods used to obtain the daily satellite data (**Figure 2**).

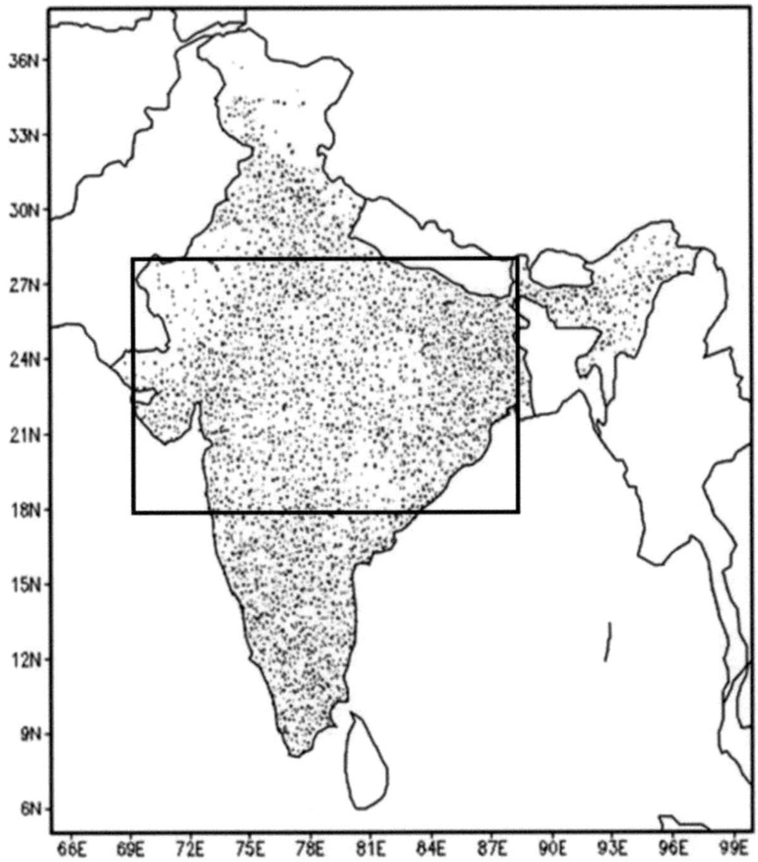

Figure 1.
Monsoon core region of India and density of rain gauges over Indian land mass (source: [20, 22]).

Algorithm	TRMM multi-satellite precipitation analysis	Integrated multi-satellite retrievals for GPM
Basic acronym	TMPA	IMERG
Data set	• 3B42Daily production multisatellite-gauge combination (precipitation) recommended for general use	• 3B-DAY.MS.MRG.3IMERG production multisatellite-gauge combination (precipitationCal) recommended for general use
Spatial grid; coverage	0.25°×0.25° lat/lon; 50°N-S	0.1°×0.1° lat/lon; 60°N-S
Current version	7	5
Time interval; span	• 3 h centered at 00, 03, ..., 21 UTC; 1998–2013 • Other value-added products in data centers	• 30 min centered at 0000, 0030, ..., 2330 UTC; 2014–present • Other value-added products in data centers
Native format	• nc4 (production) • Other value-added products in data centers	• nc4 • Other value-added products in data centers

Algorithm	TRMM multi-satellite precipitation analysis	Integrated multi-satellite retrievals for GPM

Algorithm summary	• Calibrate microwave precip rates to TRMM combined instrument
	• Merge microwave (HQ), giving preference to conical-scanners
	• Compute VAR microwave-calibrated IR precip rates
	• Fill holes in HQ merged microwave with IR estimates
	• Include gauge data by
	  • computing monthly satellite-gauge and then scaling 3 h data to sum to the monthly in each grid box (production)
	  • scaling 3 h to 3B42 with climatological coefficients (RT)
		• Calibrate microwave precip rates to GPM combined instrument
		• Merge microwave (HQ), giving preference to conical-scanners
		• Compute PERSIANN-CCS microwave-calibrated IR precip rates
		• Use CMORPH-style IR motion vectors to forward/backward propagate microwave maps, then use a Kalman filter to combine these and the IR estimates into a weighted estimate (early is forward-only)
		• Include gauge data by
		  • computing monthly satellite-gauge and then scaling 30 min data to sum to the monthly in each grid box (final)
		  • scaling 30 min to final with climatological coefficients (late and early)

Input microwave algorithms	• GPROF versions 2010v2 and 2004v for various conical scanners • NOAA MSPPS for cross-track scanners	• GPROF2014v2

Source: https://pmm.nasa.gov/resources/documents/home

Table 1.
Details on the TMPA and IMERG data sets.

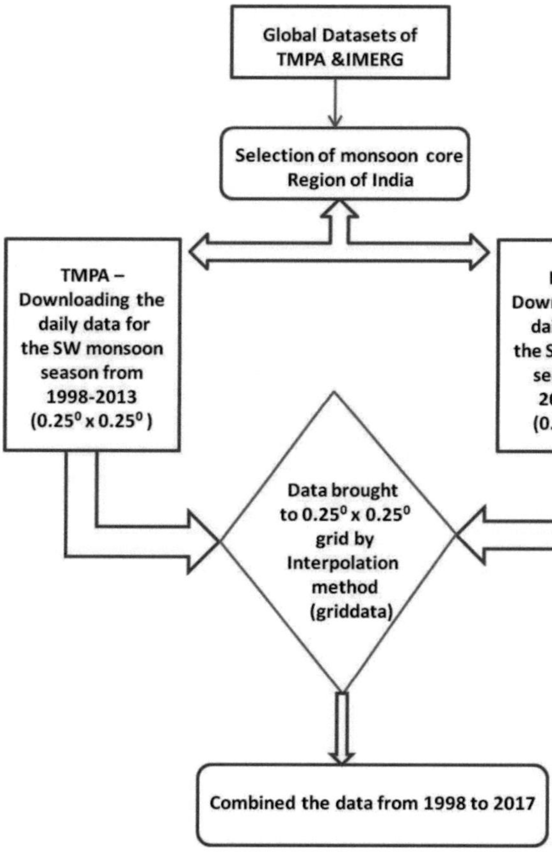

Figure 2.
Flow chart of data retrieval procedure from the satellite (TMPA and IMERG).

3. Discussion

3.1 Time series analysis of IMD and satellite (TMPA and IMERG)

Figure 3 is the daily rainfall averaged over monsoon core region for the period 1998–2017 from IMD and satellite data sets. The average daily rainfall for the season is 6.89 and 7.64 mm for IMD and satellite data sets with mean bias of 0.74 mm. Both the data sets show the monsoon characteristics such as less rainfall during the onset of monsoon, establishment of monsoon during July and August months and weakening of monsoon during the end of September. The daily variability in satellite rainfall is found to be higher than IMD indicated by the respective standard deviations (2.14 mm

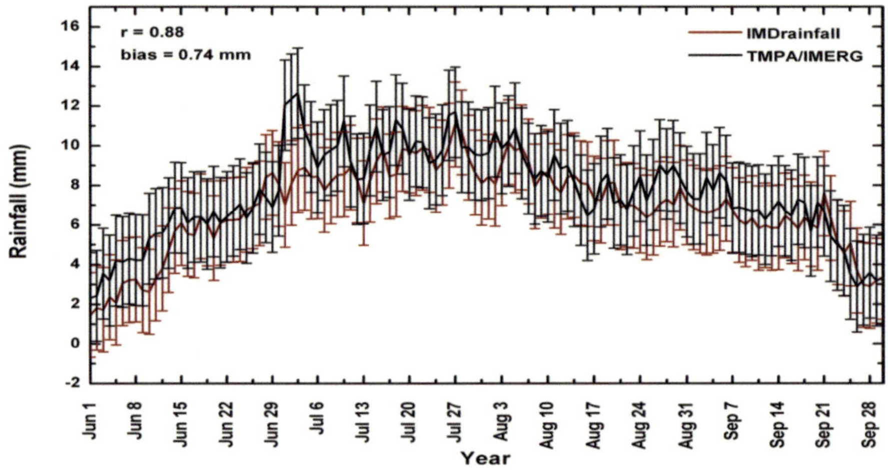

Figure 3.
Mean daily rainfall of IMD and Satellite (TMPA and IMERG) for the period 1998–2017 over monsoon core region of India during the SW monsoon season.

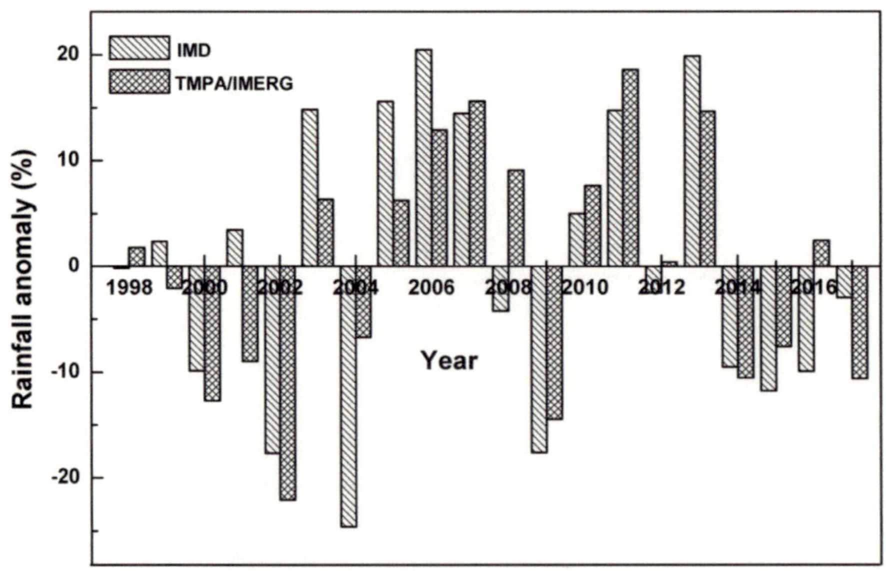

Figure 4.
Interannual variability of SW monsoon rainfall from IMD and Satellite data over monsoon core region of India.

for IMD and 2.29 mm for satellite). It is also observed that satellite is slightly over estimated the IMD rainfall. The linear association of both data sets was significant at 0.01 level with a correlation coefficient of 0.88. However, the correlation coefficient between IMD and satellite data sets varied year to year which is the indication of degree of good agreement in their magnitudes. The seasonal bias between IMD and satellite data sets over monsoon core region varied from −2.6% (2001) to a maximum of 28% (2004) which shows the less bias when whole Indian land mass is considered. The study of Uma et al. [14] on the validation of TMPA and IMD data sets report that TMPA under estimated the Indian rainfall about 100 mm for the period 1998–2007 (10 year mean rainfall for IMD and TMPA are 939 and 832 mm respectively). But over the core region, the satellite data over estimated the IMD of about 90 mm (20 year mean rainfall for IMD and satellite are 841 and 932 mm respectively). The percentage departures of IMD and satellite rainfall with respect to their long term averages are given in **Figure 4**. By and large, the satellite is able to pick up the monsoon rainfall which is evidenced by its negative anomalies during the El Niño years of 2009–2010 and 2015–2016 and positive anomalies during the La Niña year 2010–2011 respectively.

3.2 Lomb Scargle Periodogram of IMD and satellite rainfall to study ISO features

Lomb Scargle Periodogram (LSP) is a well known technique used to detect the periodic signals of given data by generating the power spectrum. The advantage of LSP is that can be applied for non-uniform and uniform data sets. When the data sets are uniform, the LSP turns to classical and in the case of non-uniform samples, it takes the Scargle generalized form of periodogram [23]. The expression for the classical periodogram can be written as

$$\mathrm{P(f)} = \frac{1}{N}\left(\sum_{n=1}^{N} gn\, e^{-2\pi i f t n}\right)^2$$

The LSP is being used to detect the periodicities of Indian rainfall. Kishore et al. [24] studied the precipitation of IMD for a period of 107 years from 1901 to 2007 using LSP and found different periodicities of 10, 15.7, 23 and 33 years with different confidence intervals. The LSP of northeast monsoon rainfall (October to December) over peninsular Indian region indicating the 20 and 30–40 days periodicity [25]. Since, the LSP is most useful in portraying the periodicities of a season; the applicability of LSP during the south west monsoon season of Indian region will give more insight to understand the dominant ISO of southwest monsoon. In the present work, we have performed the LSP for the daily rainfall data averaged over monsoon core region of India (**Figure 1**) for the period 1998–2017 for IMD as well as for satellite rainfall data sets respectively (**Figure 5**). From the **Figure 5**, we see the peaks of normalized power for different periodicities during the SW monsoon season. Wherever, the dominant periodicities are captured by IMD as well as satellite. It is well known that the dominant periodicities of Indian SW monsoon are (i) a 30–60 days oscillation, known as Madden Julian Oscillation (MJO). It is one of the tropical weather phenomena moves eastward and changes the rainfall pattern over Indian latitudes based on its arrival, (ii) oscillations having periodicity less than 20, viz., monsoon trough, tropical easterly jet, moist static stability, monsoon cloud cover, break and wet spells [26]. The overall observation of **Figure 5** suggests that the normalized power is high for the higher periodicities and is depicted by the both data sets. Two major peaks were observed during the 30–60 days MJO oscillations, one is at 40 (slightly less than) and the other is in between 50 and 60. These peaks are the representatives of strong cycles of MJO which has great impact on Indian summer monsoon. When comes to periodicities less than 20 days which are mainly known to be localized variations were

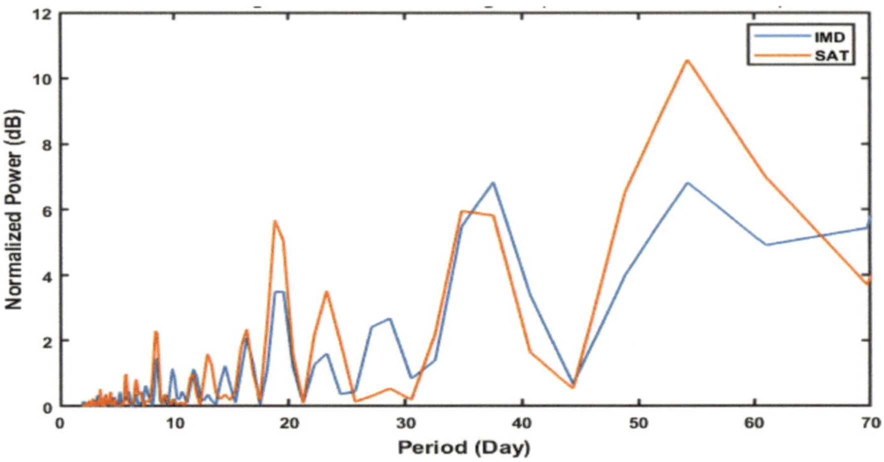

Figure 5.
LSP of daily SW monsoon rainfall of IMD and satellite for the period 1998–2017.

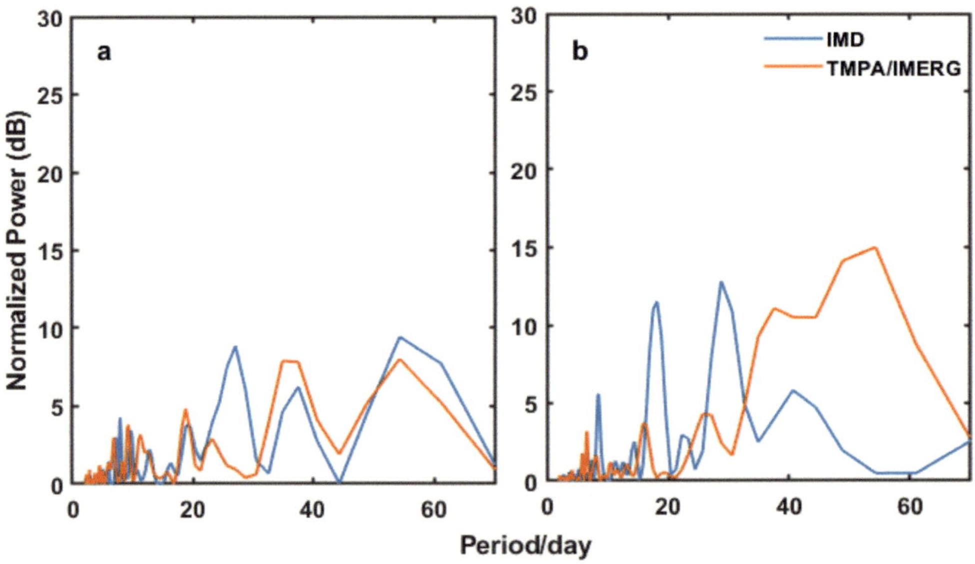

Figure 6.
(a and b) LSP of daily SW monsoon rainfall of IMD and satellite for the La Nino year 2011 and for the El Nina year 2015.

also captured by both IMD and satellites. Here we find the dominant periodicities at 20 and 7 days. However, the magnitude of normalized power during the individual years varies from satellite to IMD data sets. For this purpose, we show the LSP for the 2 years 2010 (La Niña) and 2015 (El Niño) where the monsoon core region experienced 15 and −11% of its normal seasonal rainfall (1998–2017) (**Figure 6a** and **b**). Satellite rainfall also show the similar fluctuations as IMD rainfall showed. The LSPs of 2011 and 2015 show different features though both capture the dominant periodicities. The normalized power of IMD rainfall did not vary much when compared with the satellite derived rainfall for the years 2011 and 2015. The magnitudes of normalized power of satellite show the suppressed activity of MJO in the year and 2015 and the pronounced activity during the year 2011. It is reported that the suppressed phase of MJO in the Indian Ocean causes the lower rainfall amounts when there is no deep convection [27]. Also, when the MJO is strong, higher temperature anomalies observed over lower troposphere altitudes and cooler air prevails in the upper troposphere which brings the more instability. During the presence of El Niño, the easterlies of the equatorial north Indian Ocean weakens the monsoon westerlies and in the case of La Niña, the westerly anomalies strengthen the monsoon winds. Hence, one may expect higher power of MJO in the rainfall pattern during the La Niña and lower power in the El Niño [28]. The same features were picked better by the satellite rainfall compared to IMD rainfall. As the MJO is the combination of cloud and precipitation processes, the satellite is being advantageous as it measures rainfall from the cloud properties whereas the IMD measurements are based on the ground measurements. However, as mentioned earlier, the dominant features of ISO are well captured by the satellite as the ground observations.

4. Conclusions

Main features of Indian south west monsoon were studied using the IMD and Satellite rainfall data sets. Satellite rainfall data sets show very less bias with relation to the IMD over the monsoon core region of India for the period 1998–2017. The interannual variability of satellite derived rainfall could show the impact of global teleconnections such as El Niño and La Niña evidenced by the negative and positive anomalies of rainfall from their respective means respectively. The LSP of averaged daily rainfall of IMD and satellite pick up the similar features of ISO such as Madden Julian Oscillation (30–60 days) and localized periodicities (less than 20 days). The normalized power of LSP varies from year to year from IMD to satellite rainfall, which shows the different behavior of satellite in detecting the ISO. During the El Niño and La Niña years, satellite rainfall could show better features than IMD when the normalized power is being considered. Overall, the satellite rainfall data sets over monsoon core region offer a valuable data sets with less bias and good agreement.

Acknowledgements

The second author H. A. Barbosa was funded by the Coordenação de Aperfeiçoamento de Pessoal de Nível Superior (CAPES) as part of the Pró-Alerts project (88887.091738/2014-01).

Author details

Tumuluru Venkata Lakshmi Kumar[1], Humberto Alves Barbosa[2]*, Manoj Kumar Thakur[1,3] and Franklin Paredes-Trejo[4]

1 Atmospheric Science Research Laboratory, Department of Physics, SRM Institute of Science and Technology, India

2 Laboratory for Analyzing and Processing Satellite Images, Federal University of Alagoas, Brazil

3 Tribhuvan University, Kathmandu, Nepal

4 University of the Plains Ezequiel Zamora, San Carlos, Venezuela

*Address all correspondence to: barbosa33@gmail.com

References

[1] Tao H, Fischer T, Zeng Y, Fraedrich K. Evaluation of TRMM 3B43 precipitation data for drought monitoring in Jiangsu Province, China. Water. 2016. DOI: 10.3390/w8060221

[2] Lakshmi Kumar TV, Barbosa H, Koteswara Rao K, Jothi EP. Some studies on the frequency of extreme weather events over India. Journal of Agricultural Science & Technology. 2013;**14**:1343-1356

[3] Lakshmi Kumar TV, Koteswara Rao K, Barbosa H, Uma R. Trends and extreme value analysis of rainfall pattern over homogeneous monsoon regions of India. Natural Hazards. 2014;**73**(2):1003-1017. DOI: 10.1007/s11069-014-1127-2

[4] Grimes D, Pardo-Igzquiza E, Bonifacio R. Optimal areal rainfall estimation using rain gauges and satellite data. Journal of Hydrology. 1999;**222**:93-108. DOI: 10.1016/S0022-1694(99)00092-X

[5] Novella NS, Thiaw WM. African rainfall climatology version 2 for famine early warning systems. Journal of Applied Meteorology and Climatology. 2013;**52**:588-606. DOI: 10.1175/JAMC-D-11-0238.1

[6] Chen R, Li Z, Kuligowski RJ, Ferraro R, Weng F. A study of warm rain detection using A-Train satellite data. Geophysical Research Letters. 2011;**38**:L04804. DOI: 10.1029/2010GL046217

[7] Prakash S, Gairola RM. Validation of TRMM 3B42 precipitation product over the tropical Indian Ocean using rain gauge data from the RAMA buoy array. Theoretical and Applied Climatology. 2014;**115**(3-4):451-460. DOI: 10.1007/s00704-013-0903-3

[8] Nicholson SE, Some B, Mc Collum J, Nelkin E, Klotter D, Berte Y, et al. Validation of TRMM and other rainfall estimates with a high density gauge data set for West Africa, Part II: Validation of TRMM rainfall products. Journal of Applied Meteorology. 2003;**42**:1355-1368

[9] TRMM Senior Review Proposal. In: Scott AB, Editor. Laboratory for Atmospheres. NASA Goddard Space Flight Center. 2011

[10] Thakur MK, Lakshmi Kumar TV, Dwivedi S, Narayanan MS. On the rainfall asymmetry and distribution in tropical cyclones over Bay of Bengal using TMPA and GPM rainfall products. Natural Hazards. 2018;**94**:819-832

[11] Duan Z, Bastiaanssen WGM, Liu J. Monthly and Annual Validation of TRMM Multi Satellite Precipitation Analysis (TMPA) Products in the Caspian Sea Region for the Period 1999-2003. 2012 IEEE International Geoscience and Remote Sensing Symposium; 2002. 978-1-4673-1159-5/12

[12] Nair S, Srinivasan G, Nemani R. Evaluation of multi satellite TRMM derived rainfall estimates over a western state of India. Journal of the Meteorological Society of Japan. 2009;**87**(6):927-939

[13] Narayanan MS, Shah S, Kishtawal CM, Sathiyamoorthy V, Rajeevan M, Kriplani RH. Validation of TRMM merge daily rainfall with IMD raingauge analysis over Indian land mass. Technical Report. Ahmedabad, India: Space Applications Centre; 2005

[14] Uma R, Lakshmi Kumar TV, Narayanan MS, Rajeevan M, Bhate J, Kumar KN. Assessment and spatial correspondence of TRMM and IMD gridded data sets over Indian land mass. Journal of Earth System Sciences. 2013;**122**(3):573-588

[15] Rajeevan M, Bhate J, Kale JD, Lal B. High resolution daily gridded rainfall data for the Indian region: Analysis

of break and active monsoon spells. Current Science. 2006;**91**(3):296-306

[16] Cao Y, Zhang W, Wang W. Evaluation of TRMM 3B43 data over the Yangtze River Delta of China, Scientific Reports. 2018;**8**(1):5290

[17] Xu R, Tian F, Yang L, Hu H, Lu H, Hou A. Ground validation of GPM IMERG and TRMM 3B42V7 rainfall products over southern Tibetian Plateau based on a high density rain gauge network. Journal of Geophysical Research—Atmospheres. 2017;**122**:910-924. DOI: 10.1002/2016JD025418

[18] Prakash S, Mitra AK, Aghakouchak A, Liu Z, Norouzi H, Pai DS. A preliminary assessment of GPM based multi satellite precipitation estimates over a monsoon dominated region. Journal of Hydrology. 2016;**556**:868-876. DOI: 10.1016/j/hydrol.2016.01.029

[19] He Y, Mu X, Gao P, Zhao G, Wang F, Sun W, et al. Spatial variability and periodicity of precipitation in the middle reaches of the Yellow River, China. Advances in Meteorology. 2016;**2016**:9. Article ID: 9451614. DOI: 10.1155/2016/9451614

[20] Rajeevan M, Gadgil S, Bhate J. Active and break spells of Indian summer monsoon. Journal of Earth System Science. 2010;**119**:229-247

[21] Gadgil S. The Indian Monsoon: The physics of the monsoon. Resonance. 2007;**11**:1-20

[22] Pai DS, Sridhar L, Rajeevan M, Sreejith OP, Satbhai NS, Mukhopadhyay B. Development of a new high spatial resolution (0.25 × 0.25) long period (1901-2010) daily gridded rainfall data set over India and its comparison with existing data sets over the region. Mausam. 2014;**65**(1):1-18

[23] Scargle JD. Studies in astronomical time series analysis II. Statistical aspects of special analysis of unevenly spaced data. The Astrophysical Journal. 1982;**263**:835-853

[24] Kishore P, Jothi S, Bahsha G, Rao SVB, Rajeevan M, Velicogna I, et al. Precipitation climatology over India: Validation with observations and reanalysis data sets and spatial trends. Climate Dynamics. 2015;**46**(1-2):541-556. DOI: 10.1007/s00382-015-2597-y

[25] Rajeevan M, Unnikrishnan CK, Bhate J, Niranjan Kumar K, Sreekala PP. Northeast monsoon over India: Variability and prediction. Meteorological Applications. 2012;**19**:226-236

[26] Krishnamurti TN, Bhalme HN. Oscillations of a monsoon system, Part 1: Observational aspects. Journal of the Atmospheric Sciences. 1976;**33**:1937-1954

[27] Benedict JJ, Randall DA. Observed characteristics of the MJO relative to maximum rainfall. Journal of the Atmospheric Sciences. 2007;**64**:2332-2354

[28] Achuthavarier D, Krihsnamurthy V, Kirtman BP, Huang B. Role of the Indian Ocean in the ENSO—Indian Summer monsoon teleconnection in the NCEP climate forecast system. Journal of Climate. 2012;**25**:2490-2508

Section 6

Multichannel Satellite Data Application for Global Earth Study

Chapter 8

Near- and Middle-Infrared Monitoring of Burned Areas from Space

Carlos C. DaCamara, Renata Libonati, Miguel M. Pinto and Alexandra Hurduc

Abstract

We describe a methodology to discriminate burned areas and date burning events that use a burn-sensitive (V, W) index system defined in near-/mid-infrared space. Discrimination of burned areas relies on a monthly composite of minimum of W and on the difference between this composite and that of the previous month. The rationale is to identify pixels with high confidence of having burned and aggregate new burned pixels on a contextual basis. Dating of burning events is based on the analysis of time series of *W*, and searching for the day before maximum temporal separability is achieved. The procedure is applied to the fire of Monchique, a large event that took place in the southwest of Portugal in August 2018. When the obtained pattern of burned pixels is compared against a reference map, the overall accuracy is larger than 99%; the commission and omission errors are lower than 5 and 10%, respectively; and the bias and the Dice coefficient are above 0.95 and 0.9, respectively. Differences between estimated dates of burning and reference dates derived from remote-sensed observations of active fires show a bias of 0.03 day and a root mean square difference of 0.24 day.

Keywords: burned area, dates of burning, (V, W) index system, VIIRS sensor, Monchique fire (Portugal)

1. Introduction

Vegetation fires have significant direct and indirect impacts on all components of the Earth system, including the anthroposphere. They are a source of greenhouse gases, aerosols, and trace gases to the atmosphere [1–3]; they induce modifications in most radiative forcing terms [4, 5] and disturb the radiative budget and cloud microphysics [6, 7]; they lead to changes in soil properties [8] and in the hydrological cycle [9–11]; they play a key role in biodiversity reduction, loss of genetic diversity, forest ecosystem functioning [12, 13], and land use/cover dynamics [14–16]; and they cause damages to human health [17, 18] and have adverse effects on public health and economy [19].

A thorough understanding of spatial and temporal patterns of burned area (BA) by wildfires is therefore of fundamental importance when assessing either climate or anthropogenic influences on the Earth system [20, 21]; when addressing a very wide range of subjects that include the fields of atmospheric physics and chemistry, ecology, agriculture and forestry, hydrology, biology, sociology, and economy; and when defining climate, environment, and health policies [22–26]. When specifically focusing on fire management that comprises fire prevention, fire presuppression, and fire suppression measures, reliable information about the extent, location, and time of occurrence of BA is of high added value [25]. Accurate BA information is also crucial to land and fire decision-makers, as well as to research groups and ecologists, government agencies, and NGOs when implementing environmental policies aiming to reduce socioeconomic impacts from vegetation fires on ecosystems and people [27].

The use of remotely sensed information for BA detection is well established, and there is a consensus about its usefulness from global down to regional levels [28–32]. Spaceborne sensors are a cost-effective way to map vegetation fires and the unique source of information for large areas with limited access at regional and global scales and for continuous monitoring over time [33, 34]. Over the last decades, several initiatives have been carried out to generate global and regional long-term maps of BA using remote sensing. These include, among others, (1) the 1-km L3JRC product, covering the period from April 2000 to March 2007, produced from SPOT VEGETATION data [35]; (2) the 1-km GLOBCARBON BA product, spanning the period April 1998–December 2007, derived from SPOT VEGETATION, Along Track Scanning Radiometer (ATSR-2), and Advanced ATSR (AATSR) imagery using a combination of mapping algorithms [36]; (3) the MCD45 [37] and MCD64A1 [30] BA products derived by NASA using data collected by the Moderate Resolution Imaging Spectroradiometer (MODIS); (4) the Global Fire Emissions Database (GEFD) initiative that consists in monthly BA estimates aggregated at 0.5° spatial resolution, covering the period from July 1996 to mid-2009 using four satellite data sets [38]; (5) the AQM-MODIS product [39] that was derived for Brazil and consists in monthly maps of BA at 1 km spatial resolution from 2000 up to the present; (6) the global burned area algorithm based on Medium Resolution Imaging Spectrometer (MERIS) reflectance and MODIS hotspots from 2006 to 2008 [29]; and (7) the recent global burned area product based on MODIS bands with a spatial resolution of 250 m [40].

Remote-sensed detection of burned vegetation makes use of spectral bands that are sensitive to spectral changes induced by fire events [41], namely, those associated to the deposit of char and ash on the surface and the change or destruction of vegetation structure [33]. Spectral indices have revealed to be the most appropriate to uncover changes in the radiometric signals of surfaces in operational applications [42], and a large variety of spectral indices for burned area discrimination have been developed in the last decades using a variety of techniques and different spectral bands, such as the red (R, about 0.6–0.7 μm), the near infrared (NIR, about 0.7–1.3 μm), the shorter short-wave infrared (SSWIR, about 1.3–1.9 μm), and the longer short-wave infrared (LSWIR, about 1.9–2.5 μm). Developed approaches include, among others, the Burned Area Index (BAI) [43] based on R and NIR and its improved version BAIM [44] based on NIR and LSWIR, the NIR and LSWIR-based Normalized Burn Index (NBI) [45], the Normalized Burn Ratio (NBR) [46] and derived indices from the latter [47–51], and the Mid-Infrared Burned Index (MIRBI) [52] based on SSWIR and LSWIR.

A burn-sensitive vegetation index system, the so-called (V, W) system, has also been defined on the NIR/MIR space with the aim of optimally discriminating burned vegetation [53, 54]. Here we present and discuss the use of the (V, W) index system to design an automated algorithm aiming at both mapping burned area and dating the associated burning events. As an example of application, the procedure is applied to the fire of Monchique, a large event that took place in the southwest of Portugal in August 2018 (**Figure 1**).

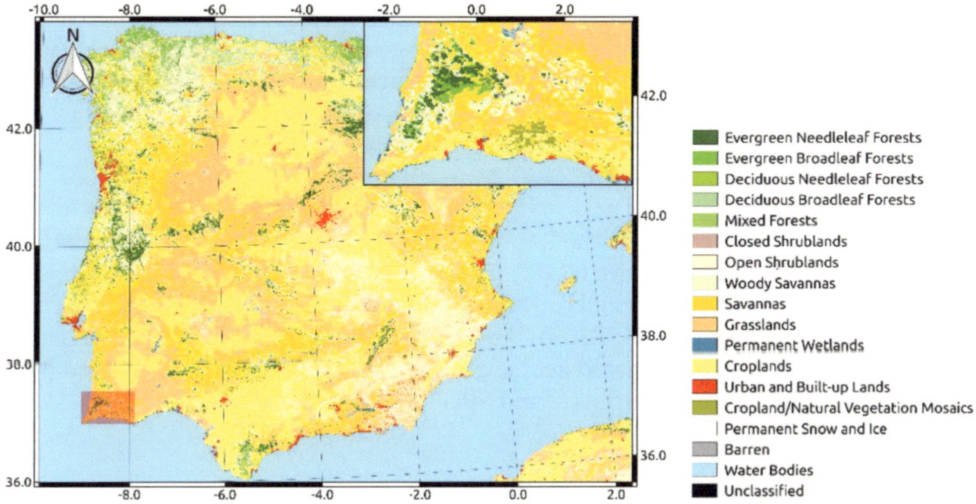

Figure 1.
Land cover map of the Iberian Peninsula showing the geographical location (shaded rectangular area) and a zoom (top right box) of the study area near the southern coast of Portugal (source of land cover data: Modis collection 6 global land cover, https://lpdaac.usgs.gov/sites/default/files/public/product_documentation/mcd12_user_guide_v6.pdf).

The fire of Monchique started on August 3 about noon and was not dominated until August 9. The fire resulted in about 27,000 hectares of burned area, 41 people injured and millions of euros in economic losses. By the second day of the event, about 700 firefighters and 11 aerial resources were fighting the fire, and this number kept increasing up to about 1400 firefighters and 14 aerial resources. The fire occurred within a context of very high temperatures and intense and highly variable winds in terrain with difficult access and high accumulation of biomass.

2. Data and pre-processing

Input data to the algorithms to compute (V, W) consist of top-of-the-atmosphere (TOA) values of middle-infrared (MIR) and thermal-infrared (TIR) radiances and of near-infrared (NIR) reflectance, as acquired by the Visible Infrared Imaging Radiometer Suite (VIIRS) instrument on board of the joint NASA/NOAA Suomi National Polar-Orbiting Partnership (Suomi-NPP) satellite [55]. VIIRS data were reprojected onto a geographical grid of 0.0045° in latitude by 0.0059° in longitude, corresponding to about 500 m in spatial resolution. Data over Portugal, covering the period of July and August 2018, were extracted from the VIIRS/NPP Level 1B 375 m product [56] and correspond to bands I2 (NIR, centered at 0.865 μm), I4 (MIR, centered at 3.74 μm), and I5 (TIR, centered at 11.45 μm).

Geolocation data, as well as land/sea mask and solar and view angle information for each VIIRS tile, were obtained from the VIIRS geolocation product (VIIRS/NPP Imagery Resolution Terrain-Corrected Geolocation). Values of MIR reflectance were then computed using VIIRS bands I4 (MIR) and I5 (TIR) radiances [57]. All images acquired at solar zenith angles (SZA) greater than 55° were rejected, and, when more than one image was available for the same day, the image selected was the one with the lowest solar zenith angle (SZA). Images used as input to the algorithm for burned area discrimination were further restricted to those with view zenith angles (VZA) not exceeding 45° in order to prevent large distortions in pixel size [53].

Information about active fire data was obtained from the VIIRS 375 m Active Fire product [58]. Finally, radiative power data were obtained from the fire radiative power (FRP) product developed by the Land Surface Analysis Satellite Application Facility (LSA SAF); this product is derived from data acquired by the Spinning Enhanced Visible and Infrared Imager (SEVIRI) onboard Meteosat Second Generation (MSG) series of EUMETSAT geostationary satellites [59].

A reference map of burned area in the study region was derived from geospatial information provided by the Rapid Mapping products of the Copernicus Emergency Management Service (EMS) [60]. The Copernicus EMS service was activated by the Portugal National Authority for Civil Protection on August 5 at 16:11 UTC (reference code EMSR303). We used the Delineation Map provided as of August 10 that has an estimated geometric accuracy of 5 m or better, derived by visual interpretation from Sentinel-2 and SPOT satellite observations.

3. Methods

3.1 Simplified (V, W)

Specially designed to discriminate burned areas, the (V, W) burn-sensitive vegetation index system is defined in a transformed MIR/NIR space that allows enhancing the spectral information about burned vegetation [53]. The transformed space is framed by the following two coordinates: (1) the distance, η, of each point in MIR/NIR space to a predefined convergence point, representative of a given target (e.g., a totally burned surface) and (2) the difference, ξ, between the respective MIR and NIR reflectance of each point. The coordinates η and ξ are accordingly defined as

$$\eta = \sqrt{(\rho_{MIR} - \rho^{0}{}_{MIR})^2 + (\rho_{NIR} - \rho^{0}{}_{NIR})^2} \tag{1}$$

$$\xi = \rho_{MIR} - \rho_{NIR} \tag{2}$$

where ρ_{MIR} and ρ_{NIR} represent values of reflectance in MIR and NIR and $(\rho^{0}_{MIR}, \rho^{0}_{NIR})$ are the coordinates of an ideally totally burned pixel.

Values of ρ^{0}_{MIR} and ρ^{0}_{NIR} for a given sensor may be estimated by the upper (lower) bound of reflectivity in MIR (NIR) for a large sample of recently burned pixels. To estimate these values for the VIIRS sensor, we used a sample of burned regions for several fires in central Portugal that occurred in 2017. Obtained estimates are $\rho^{0}_{MIR} = 0.29$ and $\rho^{0}_{NIR} = 0.06$.

The coordinate system (V, W) is then defined in the MIR/NIR space such that the following properties are met: (1) the V coordinate has a very small dispersion for pixels associated to surfaces containing organic matter and (2) the W coordinate increases with increasing water content of vegetated surfaces. Burned vegetation is characterized by very low values of *W* and by a sharp decrease of *W* following a fire event [53], both characteristics being especially conspicuous in monthly minimum composites of W and of differences of W between a given month and the previous one (**Figure 2**). In turn, non-vegetated surfaces like clouds and water bodies are characterized by low values of V.

Unlike VI3 [57] and GEMI3 [33], the (V, W) index system has the advantage of not having been heuristically derived; however, unlike traditional indices that rely on simple algebraic expressions and are easy to implement by users, the

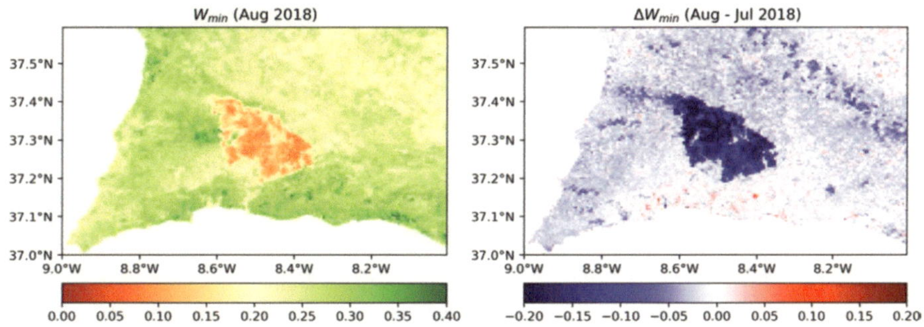

Figure 2.
Spatial distribution over the study area of the minimum composite of W_{min} for August 2018 (left panel) and of the difference of minimum composites of W_{min} between August and July 2018 (right panel).

computation of (V, W) is laborious, involving iterative methods and numerical computation of line integrals [53]. This disadvantage is circumvented by using the following approximation that is valid in a subdomain of the MIR/NIR space where the majority of observed values are located [54]:

$$V = \frac{(0.16 - 0.71\xi)}{\eta} \tag{3}$$

$$W = 1.1\eta \tag{4}$$

3.2 Discrimination of burned areas

Discrimination of burned areas for a given month is based on a procedure that uses as inputs a monthly composite of minimum of W and the difference between this minimum composite and that of the previous month together with locations of all identified hotspots during the considered month [39].

The rationale is first to identify burned pixels with high confidence of being burned and then use these points as seeds in a growing algorithm that will identify other burned pixels on a contextual basis and aggregate them as new seed points. Several studies [39, 61] have pointed out that the vast majority of hotspots are located inside or in the neighborhood of a burned area and that the number of burned pixels that are not close to a hotspot is low.

As suggested by results shown in **Figure 2**, the first seed points are therefore pixels characterized by (1) a low value of the monthly minimum composite of index W and (2) a sharp decrease in that minimum compared to the previous month.

Burned pixels are also expected to be outliers in respect to the statistical distribution pixels where no hotspots were identified. Commonly used in classification problems, the Mahalanobis distance is a measure of the distance of a point to a given distribution in units of the standard deviation in the direction to the point to the mean [62]. The square of the Mahalanobis distance in a p-dimensional space has a chi-square distribution with p degrees of freedom, a result that may be used to find outliers in a dataset [63].

Identification of burned pixels is accordingly performed in the following three steps:

- First step: Let $W\,min$ and $\Delta\,W\,min$ be the values for a given pixel of the monthly composite of minimum W and of the difference between this composite and that of the previous month; the pixel is considered as burned if all three following conditions are met:

- $W_{min} \le T1$, where $T1$ is a predefined threshold.
- $\Delta W_{min} \le T2$, where $T2$ is a predefined threshold.
- The pixel lies outside the ellipse defined in the two-dimensional space (W_{min}, ΔW_{min}) that corresponds to a predefined percentile of the Mahalanobis distance computed using pixels where no hotspots were identified; given that only pixels in the quadrant with lower $W min$ and $\Delta W min$ should be considered as burned, a given pixel is considered as burned if the Mahalanobis distance is above the predefined percentile (e.g., percentile 95) and values of $W min$ and $\Delta W min$ are sufficiently low, that is, below another predefined percentile (e.g., percentile 10, for both quantities).

- Second step: Let each pixel classified as burned in the previous step be considered as a seed point. For each seed point, a 5 × 5 buffer matrix is defined centered on it, and let N be the number of seed points inside it. If $N \geq 3$, let $\hat{W}_{min}$ and δW_{min} be the mean and the mean absolute deviation of W for these pixels. Let W^*_{min} and ΔW^*_{min} be the values of W_{min} and ΔW_{min} for a pixel inside the 5 × 5 buffer matrix that is not a seed point. This pixel is classified as burned and becomes a new seed point if all two following conditions are met:
  - $\Delta W min * < 0$
  - $\Delta W min * \le \hat{W}_{min} + \delta W_{min}$
- Third step: The previous step is repeated until no more seed points are generated.

3.3 Dating burned events

For each burned pixel identified by the algorithm above-described, the date of burning is estimated by analyzing the time series of W for that pixel and searching for the day where maximum temporal separability is achieved [64]. For most cases, time series of W present daily fluctuations of rather small amplitude (**Figure 3**) which allows identifying the day when the burning event took place by the significant decrease in W following the event. The day of burning may accordingly be identified as the one that maximizes the following index of temporal separability [65]:

$$S = \frac{2(\mu_b - \mu_a)}{\sigma_a + \sigma_b} \qquad (5)$$

where $\mu_a(\sigma_a)$ are the values of the mean (standard deviation) of index W of that pixel for a pre-specified number k of images starting at a given instant in time and $\mu_b(\sigma_b)$ are the respective values for the same k number of images before that instant in time. The time series of W is scanned by two juxtaposed windows of fixed length k, and index S is computed for every available day (**Figure 3**). The burning event is considered to have taken place in the day prior to the date when S is maximized.

3.4 Validation procedures

3.4.1 Discrimination of burned areas

The Monchique BA was validated against the data obtained from the Copernicus Emergency Management Service (EMSR303) that is used as the reference map. The

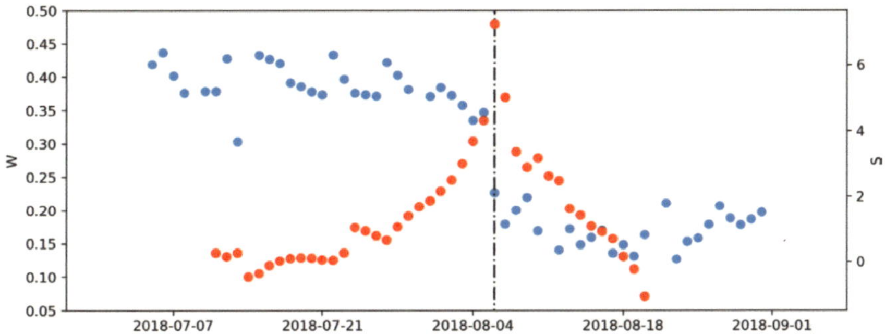

Figure 3.
Time series of indices W (blue dots, left vertical scale) and S (red dots, right vertical scale) for a pixel located inside the burned scar. The vertical black dashed line indicates the day of maximum S (Eq. (5)).

Classification map	Reference map		
	Burned	**Unburned**	
Burned	a	b	a + b
Unburned	c	d	c + d
	a + c	b + d	a + b + c + d

Table 1.
Contingency table for pixels classified as burned versus unburned.

OA	OE	CE	B	DC
$\frac{a+d}{a+b+c+d}$	$\frac{c}{a+c}$	$\frac{b}{a+d}$	$\frac{a+b}{a+c}$	$\frac{2a}{2a+b+c}$

Table 2.
Accuracy (OA), omission error (OE), commission error (CE), bias (B), and dice coefficient (DC), with a, b, c, and d as defined in **Table 1**.

quality of the classification map was assessed based on five verification measures derived from contingency tables [39]: overall accuracy (OA), omission error (OE), commission error (CE), bias (B), and Dice coefficient (DC). These verification measures are defined in **Table 2**. The agreement between the BA scar and the reference map is measured by the OA, a high value of OA reflecting a high accuracy in the classification. OE and CE are used to assess the discriminative power of the classifier. The bias should be close to one when burning events are not overestimated/underestimated. Finally, DC measures the similarity between the reference and the classification maps by overlapping the classified burned pixels to the "truly burned" pixels in the reference map.

Since the reference map has a higher resolution than the classification map, the former was projected onto the 500 m resolution grid of the latter by computing the burned fraction inside each coarser pixel. The pixel was then considered as burned if the fraction of burned area was greater than 0.5.

3.4.2 Dating burned events

Validation of estimated dates of burning was made against data of radiative power from the FRP product developed by the LSA SAF [59]. This product, together with three other active fire products derived from SEVIRI imagery, was compared against active fire data collected by the MODIS sensor, and results obtained showed a higher detection rate of active fire pixels than the other products [66]. Albeit presenting a coarser resolution of about 4 km in the study region, the repeat cycle of 15 min by the SEVIRI instrument allows for a much better temporal resolution than when comparing against VIIRS or MODIS active fires that have only two samples per day. Furthermore, the VIIRS active fires at 375 m resolution were already used in the algorithm to discriminate burned areas and therefore should not be used for validation purposes. The estimated date of each pixel classified as burned was compared to the date of observation of the nearest SEVIRI pixel where a hotspot was identified. Obtained differences between the dates of the burning of the classified burned pixels and the dates of hotspots identified by the SEVIRI instrument were then used to assess the performance of the dating methodology.

4. Example of application

The above-described procedure was applied to the study region in the southwest of Portugal in order to discriminate burned pixels during the Monchique fire episode and then estimate the respective date of burning.

As described in Sections 3.1 and 3.2, the identification of burned areas in the study region relies on monthly minimum composites of W for August (**Figure 2**, left panel) and of differences between the minimum composite of August and that of July (**Figure 2**, right panel), hereby referred to as W_{min} and ΔW_{min}, respectively. Both composites were obtained from daily values of w as derived from reflectance values of MIR and NIR from all available VIIRS images with SZA not exceeding 55° and VZA not exceeding 45°.

When values of W_{min} and ΔW_{min} for all pixels over the study region are represented in a scatter plot (**Figure 4**), two clusters may be identified: (1) one that is formed by a dense cloud with a large number of points that mostly spread over the subarea of the plot that is lower bounded by percentile 10 of the distribution of W_{min} (identified in the plot by the orange-dashed horizontal line) and left bounded by percentile 10 of the distribution of ΔW_{min} (identified by the orange-dashed vertical line) and (2) a second cluster that is composed of a less dense cloud with a lower number of points that occupy the subarea that is upper bounded by percentile 10 of the distribution of W_{min} and right bounded by percentile 10 of the distribution of ΔW_{min}.

The second cluster, formed by points with low values of both W_{min} and ΔW_{min}, is therefore likely to be associated to burned pixels. Moreover, also as to be expected in case of burned surfaces, the second cluster contains a very large fraction of pixels where hotspots were identified from the VIIRS Active Fire product (plotted as red dots). However, there are points (plotted as green dots) in the second cluster that are not associated to any hotspot, and there are also points in the first cluster that are associated to a hotspot, despite the fact that the large values of both W_{min} and ΔW_{min} are not consistent with the characteristic signature of a burned pixel. Both situations are to be expected, since (1) a pixel may burn with no active fire having been spotted by VIIRS (e.g., because of cloud or smoke screening, or because the burning took place between passages of the satellite) and (2) an identified active fire may have originated a burned area that represents a small fraction of the area of the pixel, and therefore the radiometric signature is not strong enough to be detected. Both difficulties may be circumvented in part by selecting a set of pixels with high confidence of being burned as seed points to feed into a growing algorithm.

As discussed in Section 3.2 (first step of the algorithm), seed points are defined as pixels belonging to a region of the space (W_{min},ΔW_{min}) where there is a high

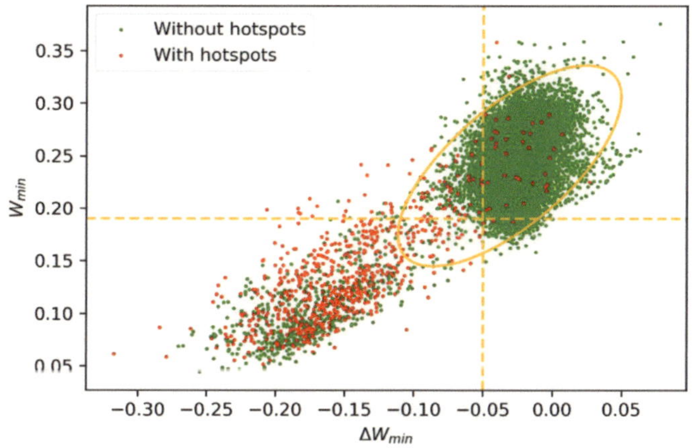

Figure 4.
W_{min} (August 2018) versus ΔW_{min} (difference between August and July 2018). Red (green) dots indicate pixels with (without) hotspots associated. The orange ellipse represents percentile 95 of the Mahalanobis distance, and the horizontal (vertical) orange-dashed line represents percentile 10 of the distribution of W_{min} (ΔW_{min}).

confidence that points are associated to burned pixels. Taking into account the above-discussed features presented by the distribution of points in the scatter plot (**Figure 4**), seed points were defined according to the following criteria:

- W_{min} < percentile 10 of W_{min}.
- ΔW_{min} < percentile 10 of ΔW_{min}.
- Points ($W min, \Delta W min$) must lie outside the ellipse representing percentile 95 of the Mahalanobis distance computed with all pixels not associated to any hotspot.

Once seed points were identified, new burned pixels were then iteratively aggregated following the procedure described in Section 3.2 (second and third steps of the algorithm).

Results obtained are shown in **Figure 5** that also provides a comparison with the reference map that was obtained from information derived from the Copernicus EMS (EMSR303). There is an overall agreement between the downscaled higher-resolution reference map and the map generated by the proposed algorithm. Deviations from the reference map, either in the form of commission or omission errors, are located along the borders of the scar and are likely to be due to small errors in geolocation or of partially burned pixels that were differently classified (as burned or unburned) by the proposed algorithm and the downscaled reference map.

The overall quality of the proposed algorithm in discriminating the burned pixels associated to the Monchique fire episode reflects on values of the contingency table that compares results from the proposed algorithm with those from the reference map from Copernicus EMS (**Table 3**) as well as on the five verification measures derived from the obtained contingency table (**Table 4**). The number of commission errors (45) and the number of omission errors (94) are one order of magnitude lower than the number of match ups (979). In turn, the overall accuracy is larger than 99%, the commission error is lower than 5%, and the omission error is lower than 10%; the bias is above 0.95, and the Dice coefficient is above 0.9.

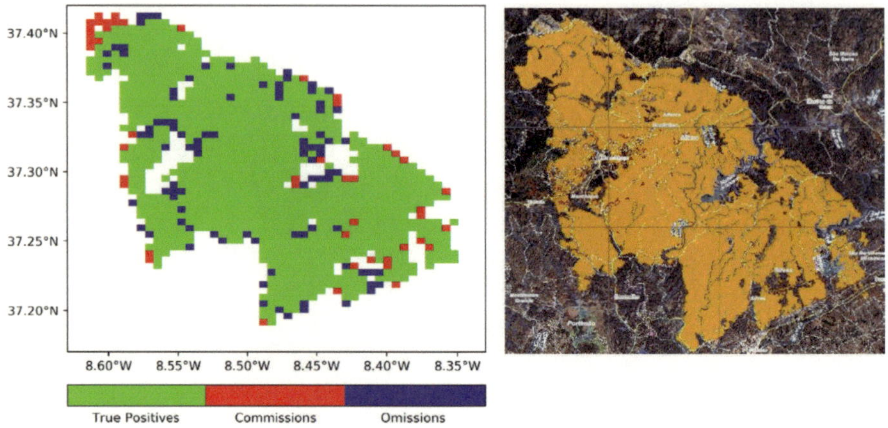

Figure 5.
Burned pixels (left panel) from the proposed algorithm and reference map (right panel) from Copernicus EMS (EMSR303). True positives, commission errors, and omission errors are colored in green, red, and blue, respectively.

BA scar	Reference map		
	Burned	**Unburned**	
Burned	979	45	1026
Unburned	94	21,357	21,451
	1073	21,404	22,477

Table 3.
As in **Table 1** *but with values obtained for the scar that resulted from the Monchique fire event of August 2018.*

OA	OE	CE	B	DC
99.4%	8.8%	4.6%	0.96	0.93

Table 4.
As in **Table 2** *but with the metrics derived from* **Table 3**.

Following the procedure described in Section 3.3, estimates were obtained of the date of burning for all pixels that were classified as burned within the study region. Results obtained (**Figure 6**, left panel) show a propagation from NW to SE, forming a pattern that is very similar to the one derived from the dates of detection of hotspots by the SEVIRI instrument (**Figure 6**, right panel). The agreement between the latter dates and the estimates by the proposed dating algorithm reflects on the obtained histogram of differences that has the null value of differences as the modal frequency, closely followed by a delay of 1 day in the estimates, such that about 70% of the pixels classified as burned have differences in the dates of less than ±1 day. When considering the distribution of differences as a whole, there is a bias of −0.03 day and a root mean square difference of 0.24 day, both values pointing out the very good overall agreement between estimates from the proposed algorithm and the reference dates derived from SEVIRI (**Figure 7**).

Results obtained using a similar procedure over the whole territory of Portugal for August and September 2005, one of the worst severe years in terms of burned area, [64] present an overall accuracy of 95.6% and commission and omission errors of 66.5 and 37.1%, respectively. However, the study encompasses a period of

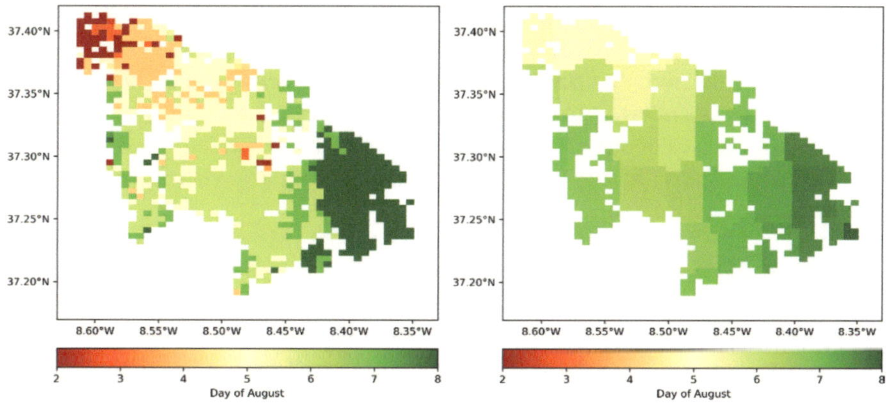

Figure 6.
Dates of burning as obtained from the proposed dating algorithm (left panel) and as derived from dates of observation of hotspots by the SEVIRI instrument (right panel).

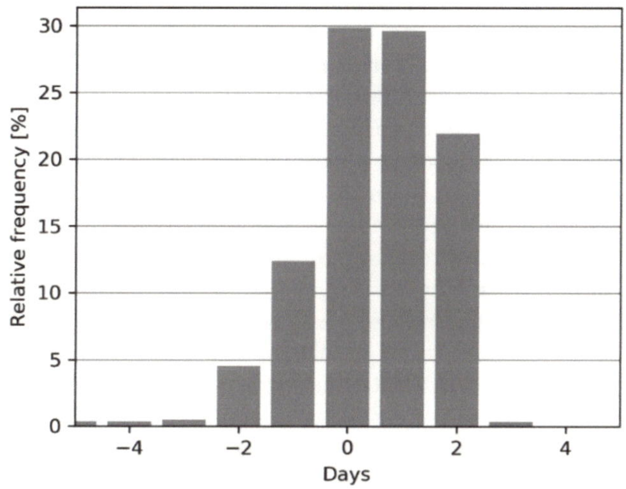

Figure 7.
Histogram of difference between dates assigned by the proposed methodology and dates derived from hotspots identified by the SEVIRI instrument.

2 months and a much wider area, covering a very large number of scars, and not a single one as in the present study. Regarding the estimated days of burning, 75% of estimated dates in the same study [64] presented deviations less than ±5 days from dates derived from hotspots identified by MODIS.

5. Conclusions

Using TOA values of MIR and TIR radiances and NIR reflectance from VIIRS 375m imagery, a set of optimal indices, V and W, were used to discriminate burned areas and to assign dates to every burned pixel. The ability of v to discriminate between vegetated and non-vegetated surfaces may be used to build up composites of w free from contamination by clouds, whereas the low values of W associated to burned surfaces suggest generating composites of minimum values of W to discriminate burned areas. Adopting this rationale, and in line with previous work [39, 64], discrimination of burned areas was performed using values (w_{min}) of a monthly composite of minimum of W and values (Δw_{min}) of differences between that composite and the one of the previous month. First seed points are identified as the pixels that (1) are outliers in respect to pixel where no hotspots were identified, (2) present low values of W_{min} characteristic of burning event, and (3) are associated to negative values of ΔW_{min}, indicating a decrease of W_{min} that is expected to occur after a burning event. New burned pixels are then successively aggregated using a seeded region-growing algorithm that starts with the previously identified seed points.

The algorithm was applied to the Monchique fire episode, a large event that occurred in southwestern Portugal during August 2018. The discriminative power of the algorithm was validated against the scar identified by Copernicus EMS303. Results obtained show that the (V, W) algorithm is suited to discriminate burned area over a mainland Portugal, supported by the good agreement, with a Dice coefficient of 0.933, between the burned area scar and the reference map. The commission and omission errors have values of 9 and 5%, respectively. Estimated dates of burning, obtained through analysis of time series of values of W, were compared against times of observation of hotspots obtained from the SEVIRI FRP product. About 70% of the estimated dates presented deviations of 1 day or less.

The development of reliable algorithms to discriminate and date burned areas is crucial for a better understanding of the biosphere-atmosphere interactions, for estimating burning emissions, for future projections of fire regime, and for mitigation and adaptation actions in Portugal, which is recurrently affected by severe fire events. In particular, accurate estimates of the date of burning are crucial when considering fire regime modeling, due to the constraint imposed by biomass availability into the spread of fire, and are also important for reducing uncertainties in biomass burning emissions [34]. The recent VIIRS sensor will allow the development of new burned area products at high spatial resolution, continuing and enhancing the imaging of the Earth initiated by the Advanced Very High-Resolution Radiometer (AVHRR) and the MODIS instruments. The present work represents a first attempt to assess the potential of using VIIRS imagery to identify burnt scars in Portugal. Results obtained in this work and in related previous ones pave the way to the generation of a long-term series of burned area maps containing accurate information about the extent, location, and time of occurrence of vegetation fires.

Acknowledgements

This research is supported by FAPESP/FCT Project Brazilian Fire-Land-Atmosphere System (BrFLAS) (FCT 2015/01389-4 and FAPESP/1389/2014) and by the EUMETSAT Land Surface Analysis Satellite Application Facility (LSA SAF). Research by Renata Libonati was funded by Serrapilheira Institute (grant number Serra-1708-15159) and supported by Centro de Estudos Florestais (CEF) of the University of Lisbon, a research unit funded by FCT (UID/AGR/00239/2013). Research by Miguel M. Pinto was supported by FCT through PhD grant PD/BD/142779/2018. Research by Alexandra Hurduc was supported by a grant in the framework of Project "Reabilitação das Áreas Queimadas na Freguesia de Alvares," financed by donation of Observador on time SA.

Conflict of interest

The authors declare they have no conflicts of interest.

Author details

Carlos C. DaCamara[1*], Renata Libonati[2,3], Miguel M. Pinto[1] and Alexandra Hurduc[1]

1 Faculdade de Ciências, Instituto Dom Luiz, Universidade de Lisboa, Lisboa, Portugal

2 Departmento de Meteorologia, Instituto de Geociências, Universidade Federal do Rio de Janeiro, Rio de Janeiro, Brazil

3 Centro de Estudos Florestais, Universidade de Lisboa, Lisboa, Portugal

*Address all correspondence to: cdcamara@fc.ul.pt

References

[1] Dentener F, Kinne S, Bond T, Boucher O, Cofala J, Generoso S, et al. Emissions of primary aerosol and precursor gases in the years 2000 and 1750 prescribed data-sets for AeroCom. Atmospheric Chemistry and Physics. 2006;**6**:4321-4344. DOI: 10.5194/acp-64321-2006

[2] Zhang J, Webster J, Powers RF, Mills J. Reforestation after the fountain fire in northern California: An untold success story. Journal of Forestry. 2008;**106**: 425-430. DOI: 10.1093/jof/106.8.425

[3] van Leeuwen W, Casady G, Neary D, Bautista S, Alloza J, Carmel J, et al. Monitoring post-wildfire vegetation response with remotely sensed time series data in Spain, USA and Israel. International Journal of Wildland Fire. 2010;**19**:75-93. DOI: 10.1071/WF08078

[4] Bowman DMJS, Balch JK, Artaxo P, Bond WJ, Carlson JM, Cochrane MA, et al. Fire in the Earth system. Science. 2009;**324**(5926):481-484. DOI: 10.1126/science.1163886

[5] Storelvmo T, Leirvik T, Lohmann U, Phillips PC, Wild M. Disentangling greenhouse warming and aerosol cooling to reveal Earth's climate sensitivity. Nature Geoscience. 2016;**9**(4):286. DOI: 10.1038/ngeo2670

[6] Andreae MO, Rosenfeld D, Artaxo P, Costa AA, Frank GP, Longo KM, et al. Smoking rain clouds over the Amazon. Science. 2004;**303**:1337-1342. DOI: 10.1126/science.1092779. DOI: 10.1126/science.1092779

[7] Lohmann U, Rotstayn L, Storelvmo T, Jones A, Menon S, Quaas J, et al. Total aerosol effect: Radiative forcing or radiative flux perturbation? Atmospheric Chemistry and Physics. 2010;**10**(7):3235-3246. DOI: 10.5194/acp-10-3235-2010

[8] Certini G. Effects of fire on properties of forest soils: A review. Oecologia. 2005;**143**(1):1-10. DOI: 10.1007/s00442-004-1788-8

[9] Rosenfeld D. TRMM observed first direct evidence of smoke from forest fires inhibiting rainfall. Geophysical Research Letters. 1999;**26**(20):3105-3108. DOI: 10.1029/1999GL006066

[10] Menon S, Hansen J, Nazarenko L, Luo Y. Climate effects of black carbon aerosols in China and India. Science. 2002;**297**:2250-2253. DOI: 10.1126/science.1075159

[11] Koren I, Kaufman YJ, Remer LA, Martins JV. Measurement of the effect of Amazon smoke on inhibition of cloud formation. Science. 2004;**303**:1342-1345. DOI: 10.1126/science.1089424

[12] Fisher B, Turner RK, Morling P. Defining and classifying ecosystem services for decision making. Ecological Economics. 2009;**68**(3):643-653. DOI: 10.1016/j.ecolecon.2008.09.014

[13] Driscoll D, Lindenmayer D, Bennett AF, Bode M, Bradstoc RA, Cary G, et al. Fire management for biodiversity conservation: Key research questions and our capacity to answer them. Biological Conservation. 2010;**143**(9):1928-1939. DOI: 10.1016/j.biocon.2010.05.026

[14] Sellers PJ, Bounoua L, Collatz GJ, Randall DA, Dazlich DA, Los SO, et al. Comparison of radiative and physiological effects of doubled atmospheric CO_2 on climate. Science. 1996;**271**:1402-1406. DOI: 10.1126/science.271.5254.1402

[15] Jin Y, Roy DP. Fire-induced albedo change and its radiative forcing at the surface in northern Australia. Geophysical Research

Letters. 2005;**32**:L13401. DOI: 10.1029/2005GL022822

[16] Lambin EF, Geist HJ. Land-Use and Land-Cover Change. Local Processes and Global Impacts. Berlin: Springer Science & Business Media; 2008. p. WP6319

[17] Bowman DMJS, Johnston FH. Wildfire smoke, fire management, and human health. EcoHealth. 2005;**2**(1):76-80. DOI: 10.1007/s10393-004-0149-8

[18] Fowler CT. Human health impacts of forest fires in the southern United States: A literature review. Journal of Ecological Anthropology. 2003;**7**(1):39-63. DOI: 10.5038/2162-4593.7.1.3

[19] Patz JA, Engelberg D, Last J. The effects of changing weather on public health. Annual Review of Public Health. 2000;**21**(1):271-307. DOI: 10.1146/annurev.publhealth.21.1.271

[20] Bowman D. Wildfire science is at a loss for comprehensive data. Nature. 2018;**560**:7. DOI: 10.1038/d41586-018-05840-4

[21] Andela N, Morton DC, Giglio L, Chen Y, Van Der Werf GR, Kasibhatla PS, et al. A human-driven decline in global burned area. Science. 2017;**356**:1356-1362. DOI: 10.1126/science.aal4108

[22] Flannigan MD, Krawchuk MA, de Groot WJ, Wotton BM, Gowman LM. Implications of changing climate for global wildland fire. International Journal of Wildland Fire. 2009;**18**(5):483-507. DOI: 10.1071/WF08187

[23] Langmann B, Duncan B, Textor C, Trentmann J, van der Werf GR. Vegetation fire emissions and their impact on air pollution and climate. Atmospheric Environment. 2009;**43**(1):107-116. DOI: 10.1016/j.atmosenv.2008.09.047

[24] Kochi I, Donovan GH, Champ PA, Loomis JB. The economic cost of adverse health effects from wildfire-smoke exposure: A review. International Journal of Wildland Fire. 2010;**19**(7):803-817. DOI: 10.1071/WF09077

[25] Bowman DM, Balch J, Artaxo P, Bond WJ, Cochrane MA, D'Antonio CM, et al. The human dimension of fire regimes on Earth. Journal of Biogeography. 2011;**38**(12):2223-2236. DOI: 10.1111/j.1365-2699.2011.02595.x

[26] Johnston FH, Henderson SB, Chen Y, Randerson JT, Marlier M, DeFries RS, et al. Estimated global mortality attributable to smoke from landscape fires. Environmental Health Perspectives. 2012;**120**(5):695. DOI: 10.1289/ehp.1104422

[27] Pacheco AP, Claro J, Fernandes PM, de Neufville R, Oliveira TM, Borges JG, et al. Cohesive fire management within an uncertain environment: A review of risk handling and decision support systems. Forest Ecology and Management. 2015;**347**:1-17. DOI: 10.1016/j.foreco.2015.02.033

[28] Nogueira J, Ruffault J, Chuvieco E, Mouillot F. Can we go beyond burned area in the assessment of global remote sensing products with fire patch metrics? Remote Sensing. 2016;**9**(1):7. DOI: 10.3390/rs9010007

[29] Alonso-Canas I, Chuvieco E. Global burned area mapping from ENVISAT-MERIS and MODIS active fire data. Remote Sensing of Environment. 2015;**163**:140-152. DOI: 10.1016/j.rse.2015.03.011

[30] Giglio L, Boschetti L, Roy DP, Humber ML, Justice CO. The collection 6 MODIS burned area mapping algorithm and product. Remote Sensing of Environment. 2018;**217**:72-85. DOI: 10.1016/j.rse.2018.08.005

[31] Pereira AA, Pereira J, Libonati R, Oom D, Setzer AW, Morelli F, et al. Burned area mapping in the Brazilian savanna using a one-class support vector machine trained by active fires. Remote Sensing. 2017;**9**(11):1161. DOI: 10.3390/rs9111161

[32] Kasischke ES, French NHF, Harrell P, Christensen NL Jr, Ustin SL, Barry D. Monitoring of wildfires in boreal forests using large area AVHRR NDVI composite image data. Remote Sensing of Environment. 1993;**45**:61-71

[33] Pereira JMC. A comparative evaluation of NOAA/AVHRR vegetation indices for burned surface detection and mapping. IEEE Transactions on Geoscience and Remote Sensing. 1999;**37**(1):217-226. DOI: 10.1109/36.739156

[34] Mouillot F, Schultz MG, Yue C, Cadule P, Tansey K, Ciais P, et al. Ten years of global burned area products from spaceborne remote sensing—A review: Analysis of user needs and recommendations for future developments. International Journal of Applied Earth Observation and Geoinformation. 2014;**26**:64-79. DOI: 10.1016/j.jag.2013.05.014

[35] Tansey K, Gregoire JM, Defourny P, Leigh R, Pekel JFO, van Bogaert E, et al. A new, global, multi-annual (2000-2007) burnt area product at 1 km resolution. Geophysical Research Letters. 2018;**35**:1. DOI: 10.1029/2007GL031567

[36] Plummer S, Arino O, Simon M, Steffen W. Establishing an earth observation product service for the terrestrial carbon community: The GLOBCARBON initiative. Mitigation and Adaptation Strategies for Global Change. 2006;**11**:97-111. DOI: 10.1007/s11027-006-1012-8

[37] Roy DP, Boschetti L, Justice CO, Ju J. The collection 5 MODIS burned area product: Global evaluation by comparison with the MODIS active fire product. Remote Sensing of Environment. 2008;**112**:3690-3707. DOI: 10.1016/j.rse.2008.05.013

[38] Giglio L, Randerson JT, van der Werf GR, Kasibhatla PS, Collatz GJ, Morton DC, et al. Assessing variability and long-term trends in burned area by merging multiple satellite fire products. Biogeosciences. 2010;**7**:1171-1186. DOI: 10.5194/bg-7-1171-2010

[39] Libonati R, DaCamara CC, Setzer AW, Morelli F, Melchiori AE. An algorithm for burned area detection in the Brazil Cerrado using 4 μm MODIS imagery. Remote Sensing. 2015;7:15782-15803. DOI: 10.3390/rs71115782

[40] Chuvieco E, Lizundia-Loiola J, Pettinari ML, Ramo R, Padilla M, Mouillot F, et al. Generation and analysis of a new global burned area product based on MODIS 250 m reflectance bands and thermal anomalies. Earth System Science Data Discussions. 2018;**512**:1-24. DOI: 10.5194/essd-2018-46

[41] Trigg S, Flasse S. Characterising the spectral-temporal response of burned savannah using in situ spectroradiometry and infrared thermometry. International Journal of Remote Sensing. 2000;**21**:3161-3168. DOI: 10.1080/01431160050145045

[42] Verstraete MM, Pinty M. Designing optimal spectral indexes for remote sensing applications. IEEE Transactions on Geoscience and Remote Sensing. 1996;(5):1254-1265. DOI: 10.1109/36.536541

[43] Isabel MDPM. Cartografía e inventario de incendios forestales en la Península Ibérica a partir de imágenes NOAA-AVHRR. Universidad de Alcalá de Henares; 1999

[44] Martín MP, Gómez I, Chuvieco E. Performance of a burned-area index

(BAIM) for mapping Mediterranean burned scars from MODIS data. In: Proceedings of the 5th International Workshop on Remote Sensing and GIS Applications to Forest Fire Management: Fire Effects Assessment. Paris: Universidad de Zaragoza, GOFC GOLD, EARSeL; 2005. pp. 193-198

[45] Alleaume S, Hely C, Le Roux J, Korontzi S, Swap RJ, Shugart HH, et al. Using MODIS to evaluate heterogeneity of biomass burning in southern African savannahs: A case study. International Journal of Remote Sensing. 2005;**26**:4219-4237. DOI: 10.1080/01431160500113492

[46] USGS. Product Guide: Landsat surface reflectance-derived spectral indices [Internet]. 2007. Available from: https://landsat.usgs.gov/sites/default/files/documents/si_product_guide.pdf [Accessed: October 28, 2018]

[47] Loboda T, O'Neal KJ, Csiszar I. Regionally adaptable dNBR-based algorithm for burned area mapping from MODIS data. Remote Sensing of Environment. 2007;**109**:429-442. DOI: 10.1016/j.rse.2007.01.017

[48] Miller JD, Yool SR. Mapping forest post-fire canopy consumption in several overstory types using multi-temporal Landsat TM and ETM data. Remote Sensing of Environment. 2002;**82**(2-3):481-496. DOI: 10.1016/S0034-4257(02)00071-8

[49] Boer MM, Macfarlane C, Norris J, Sadler RJ, Wallace J, Grierson PF. Mapping burned areas and burn severity patterns in SW Australian eucalypt forest using remotely-sensed changes in leaf area index. Remote Sensing of Environment. 2008;**112**:4358-4396. DOI: 10.1015/j.rse.2008.08.005

[50] Miller JD, Thode AE. Quantifying burn severity in a heterogeneous landscape with a relative version of the delta normalized burn ratio (dNBR). Remote Sensing of Environment. 2007;**109**:66-80. DOI: 10.1016/j.rse.2006.12.006

[51] Bastarrika A, Chuvieco E, Martín MP. Mapping burned areas from Landsat TM/ETM+ data with a two phase algorithm: Balancing omission and commission errors. Remote Sensing of Environment. 2011;**105**:1003-1012. DOI: 10.1016/j.rse.2010.12.005

[52] Trigg S, Flasse S. An evaluation of different bi-spectral spaces for discriminating burned shrub-savannah. International Journal of Remote Sensing. 2001;**22**(13):2641-2647. DOI: 10.1080/01431160110053185

[53] Libonati R, DaCamara CC, Pereira JMC, Peres LF. On a new coordinate system for improved discrimination of vegetation and burned areas using MIR/NIR information. Remote Sensing of Environment. 2011;**114**:831-843. DOI: 10.1016/j.rse.2011.02.006

[54] DaCamara CC, Libonati R, Ermida SL, Calado TJ. A user-oriented simplification of the (V, W) burn-sensitive vegetation index system. IEEE Geoscience and Remote Sensing Letters. 2016;**13**(12):1822-1826. DOI: 10.1109/LGRS.2016.2614319

[55] NASA. Data Product User Guide for Suomi-National Polar-Orbiting Partnership (S-NPP) Sounder Science Investigator-led Processing System (SIPS) Advanced Technology Microwave Sounder (ATMS) Level 1B Products [Internet]. 2017. Available from: https://docserver.gesdisc.eosdis.nasa.gov/repository/Mission/SNPP_Sounder/3.3_ScienceDataProductDocumentation/3.3.4_ProductGenerationAlgorithms/ATMS_Readme_Vers_1_20170508.pdf [Accessed: October 28, 2018]

[56] NASA. Visible Infrared Imaging Radiometer Suite (VIIRS) 375 m &

750 m Active Fire Detection Data Sets Based on NASA VIIRS Land Science Investigator Processing System (SIPS) Reprocessed Data—Version 1 [Internet]. 2017. Available from: https://lpdaac.usgs.gov/sites/default/files/public/product_documentation/vnp14_user_guide_v1.3.pdf [Accessed: October 28, 2018]

[57] Kaufman YJ, Remer LA. Detection of forests using MID-IR reflectance: An application for aerosol studies. IEEE Transactions on Geoscience and Remote Sensing. 1994;**32**:672-683. DOI: 10.1109/36.297984

[58] Schroeder W, Oliva P, Giglio L, Csiszar IA. The new VIIRS 375 m active fire detection data product: Algorithm description and initial assessment. Remote Sensing of Environment. 2014;**143**:85-96. DOI: 10.1016/j.rse.2013.12.008

[59] Wooster MJ, Roberts G, Freeborn PH, Xu W, Govaerts Y, Beeby R, et al. LSA SAF Meteosat FRP products—Part 1: Algorithms, product contents, and analysis. Atmospheric Chemistry and Physics. 2015;**15**(22):13217-13239. DOI: 10.5194/acp-15-13217-2015

[60] Copernicus. Emergency Management Service [Internet]. 2018. Available from: http://emergency.copernicus.eu/ [Accessed: October 28, 2018]

[61] Hantson S, Padilla M, Cort D, Chuvieco E. Strengths and weaknesses of MODIS hotspots to characterize global fire occurrence. Remote Sensing of Environment. 2013;**131**:152-159. DOI: 10.1016/j.rse.2012.12.004

[62] Mahalanobis PC. On the generalised distance in statistics. Proceedings of the National Institute of Sciences of India. 1936;**2**(1):49-55

[63] Wilks DS. Statistical Methods in the Atmospheric Sciences. Vol. 100. Oxford: Academic Press; 2006. 676 pp

[64] Panisset J, DaCamara CC, Libonati R, Peres LF, Calado TJ, Barros A. Assigning dates and identifying areas affected by fires in Portugal based on MODIS data. Anais da Academia Brasileira de Ciências. 2017;**89**(3):1487-1501. DOI: 10.1590/0001-3765201720160707

[65] Giglio L, Loboda T, Roy DP, Quayle B, Justice CO. An active-fire based burned area mapping algorithm for the MODIS sensor. Remote Sensing of Environment. 2009;**113**:408-420. DOI: 10.1016/j.rse.2008.10.006

[66] Roberts G, Wooster MJ, Xu W, Freeborn PH, Morcrette JJ, Jones L, et al. LSA SAF Meteosat FRP products—Part 2: Evaluation and demonstration for use in the Copernicus Atmosphere Monitoring Service (CAMS). Atmospheric Chemistry and Physics. 2015;**15**:13241-13267. DOI: 10.5194/acp-15-13241-2015

Chapter 9

The Use of Visible Geostationary Operational Meteorological Satellite Imagery in Mapping the Water Balance over Puerto Rico for Water Resource Management

John R. Mecikalski and Eric W. Harmsen

Abstract

A solar insolation satellite remote sensing product for Puerto Rico, the US Virgin Islands (USVI), Dominican Republic, Haiti, Jamaica, and Cuba became available in 2009 through a collaboration between the University of Puerto Rico-Mayagüez Campus and the University of Alabama in Huntsville. Solar insolation data are available at 1 km resolution for Puerto Rico and the USVI and 2 km resolution for the other islands, as derived from 500 m resolution GOES-16 visible imagery. The insolation data demonstrate the powerful utility of satellite-derived fields for water resource applications, specifically the routine production of potential and reference evapotranspiration. This chapter describes the theoretical background and technical approach for estimating components of the daily water and energy balance in Puerto Rico. Useful information can be obtained from the model, which benefits disaster and emergency management, agriculture, human health, ecology, coastal water management, and renewable energy development at the island scale.

Keywords: incoming solar radiation, insolation, GOES, Puerto Rico, Caribbean, evapotranspiration, remote sensing, water resource management, reference evapotranspiration, potential evapotranspiration

1. Introduction

Estimates of incoming solar radiation (also known as "insolation") have been made from geostationary satellite data for many years, since the early to mid-1970s [1]. Related to the present effort, Geostationary Operational Environmental Satellite (GOES) visible channel (~0.64 μm) data have been processed within a scalable and flexible insolation model, which is well documented and described in detail below. For ongoing water management support over Puerto Rico and the broader Caribbean, the Diak-Gautier insolation model [2] has been specifically structured to provide daily integrated, gridded solar insolation at 1–2 km spatial resolution. The insolation model has been rigorously tested and validated and operates on GOES imagery from GOES–4/–5 through the present day GOES–16/–17. Geostationary satellites are optimal for providing spatially and temporally continuous fields across all regions in their ±55° latitude field of view, which as noted is a significant advantage over the use of only ground-based instrumentation. The use of a satellite-based insolation algorithm also ensures that a consistent algorithm is applied across an entire region, one which relies on data from only one instrument, specifically, the GOES Imager.

Over Puerto Rico (PR) and the Caribbean, as well as in other subtropical and tropical regions, evapotranspiration (ET) is a critical variable for water management, both in hydrologic flow simulations involving potential ET (PET) and water allocation and agricultural water use involving reference ET (RET or ET_o). Importantly, *solar insolation is a large, yet often unknown, determinant for temporal variation in PET and RET. Solar insolation is a primary determinant of spatial variation, particularly in areas with heterogeneous cloud cover, as common to subtropical and tropical regions where small cumulus clouds dominate the regional cloud climatology*.

For an ET product to be desirable, it must be spatially continuous, rather than consisting of only point values derived from local weather station networks. Thus, mapping of ET is greatly facilitated by satellite-derived estimates that contain the actual spatial variability and distribution of solar insolation. Prior to 2009, regions across Puerto Rico and the Caribbean did not had access to a consistent, spatially continuous method of computing RET and PET. The original motivation for development of the Geostationary Operational Environmental Satellite-Water and Energy Balance (GOESWEB) model was to develop a robust insolation calibration framework coupled to a satellite-based insolation model, to provide a key radiative dataset that can grow over time toward 10-year and longer timeframes, thus forming an ET climatology that can be extended indefinitely.

The GOES-based insolation datasets are used in conjunction with other information, including net radiation (R_n), air temperature, relative humidity, wind speed, and land cover information, in the formulation of daily, 1- and 2-km estimates of RET across the Caribbean. RET is valuable for farm- and city-based water management, as well as irrigation scheduling; PET can be used as input into surface and groundwater hydrological models, whereas the solar insolation data themselves may be used as data input in certain ecosystem models.

2. GOES solar insolation data

The use of geostationary satellite visible data has been used for estimating solar insolation for over 30 years. The main methods used for such estimation range from statistical-empirical relationships, such as [3], to varying complex physical models [2, 4–12]. Studies such as [13, 14] proved the utility and feasibility of satellite-estimated solar insolation methods, demonstrating that fairly accurate results can be produced from such models; hourly insolation estimates obtained from the most current models are within 5–10% of ground-based pyranometer data, during clear-sky conditions (15–30% for all sky conditions), while daily estimates are found to be within 10–15% [15]. Studies by [16, 17] have further highlighted the overall utility of these methods.

The main advantages of using satellite-estimated insolation, over those collected by pyranometer networks, include wide-area spatial coverage, high spatial resolution (1–2 km), and the ability to produce useful data in remote, inaccessible, or in potentially hazardous areas, over large water bodies and oceans (e.g., [18]), and in locations where the installation of a ground-based pyranometer network is prohibitive. As an alternative to the methods used in this study, [19, 20] describe the use of the Global Energy and Water Cycle Experiment (GEWEX) Surface Radiation Budget (SRB) downward solar flux [11], as used within the North American Land Data

Assimilation project. Error statistics for the SRB product are comparable to those shown in [21], as used in this study, yet SRB resolutions are at best 0.5° and 3 hourly [22].

Related to the PRWEB applications to be developed here, the solar insolation product is derived from National Oceanic and Atmospheric Administration (NOAA) GOES-East satellite visible (0.64 μm) imagery. These data were processed using [4] methods to produce daily integrated solar insolation throughout Puerto Rico at 1-km horizontal spatial resolution. This 1-km resolution is chosen as it provides solar insolation observations between cumulus clouds, which comprise a significant component of the cloud climatology in subtropical regions.

2.1 Details of the GOES solar insolation model

The GOES solar insolation model is developed by [4], which was later modified by [2] and updated by [23], and most recently by [24], which is the 2017 version of the solar insolation model employed in this study. This model will be referred to as the "GD" model from this point forward and employs a simple physical model that represents cloud and atmosphere radiative processes. The GD model was shown to perform even better than more complex solar insolation methods over a variety of land-surface and climatic conditions [5, 17, 18, 23, 26]. When comparing with pyranometer data, these prior studies list root mean square errors in hourly and daily insolation estimates as a percentage of the mean pyranometer observed value, which range from 17–28% to 9–10%, respectively. In [24, 25] the higher magnitudes of these errors were reported (~28 and ~10%, respectively) in a study over northern central Florida using GOES–12 data. However, the GD model has been proven to be valuable in operational use of near-real-time, regional-, and continental-scale insolation estimates for several main applications, including land-surface carbon and water flux assessments [27–29], the generation of agricultural forecasting products [30, 31], and subsurface hydrologic modeling.

The GD model is based on conservation of radiant energy in the Earth-atmosphere column, with two modes for estimating solar insolation received at Earth's surface: (1) clear and (2) cloudy conditions. These modes are determined based on satellite-derived, visible channel surface albedo data. A reference albedo grid representative of clear-sky conditions per satellite pixel is developed within the GD algorithm, which captures the temporal changes in land-surface characteristics over time and season. This running 2-week minimum of this albedo data, reassessed at solar noon daily, is stored for each GOES satellite visible data pixel. This approach is considered representative of the true land-surface albedo, which is more accurate than using the daily estimated value as the latter may be corrupted by high albedo values when even low-cloud amounts are present during a given day. Note that this minimum albedo is wavelength-specific, is unique to the GOES Imager visible sensor (which includes some near-infrared contribution), and is not a true surface albedo.

As the GD algorithm runs across a series of GOES images per day, the digital brightness at each image pixel is compared to that of the stored clear-sky reference 2-week minimum albedo for that pixel. If the brightness exceeds that threshold, the pixel is deemed partly or completely cloudy. Based on this determination per GOES pixel, either the clear or cloudy model of atmospheric radiation processes (within the GD model) is used to calculate surface solar insolation received. Both clear and cloudy models incorporate parameterizations for Raleigh scattering, ozone absorption, and water vapor absorption within the atmospheric column, using simple bulk relationships, such as fixed ozone and aerosol contents. This rough parameterization works because these produce secondary sources of error to the instantaneous surface solar insolation. The cloudy GD component estimates a cloud-top albedo and separately accounts for atmospheric effects above and below the cloud.

For the water vapor absorption parameterization, a fixed, approximate annual median value of precipitable water (PW) of 3.0 cm was used, which is considered appropriate for Puerto Rico. This annual median value helps to estimate atmospheric column-integrated PW during the initial processing. [PW is defined as the amount of water that would precipitate out of a vertical column of the atmosphere if all the water vapor were condensed into liquid]. PW data are used to calculate the slantwise path and subsequently the absorption coefficients [4]. Real-time PW data from numerical forecast model output may also be used in the GD model, versus setting a constant value.

2.2 GOES data processing and quality control

The GOES-East series of satellites (the most recent additions being GOES–13 and –16) are in geostationary orbit above the Earth's equator at −75° W, which provides continuous, 5–15-minute resolution observations in visible and infrared radiation channels at high spatial (500 m to 1 km). GOES data are thus ideal for high-resolution estimates of solar insolation as used in GOESWEB, to be described below. Although the GOES visible sensors have a nadir (the point directly below the satellite) spatial resolution of 1 km (GOES–13 and prior) or 500 m (GOES–16), this resolution decreases the further from nadir the instrument scans: for Puerto Rico, the highest resolution attainable is about 1.25 km and 525 m, respectively, for GOES–13 and –16. All solar insolation data used for this study were provided at 1-km resolution. A simple method for computing sunrise and sunset times per pixel across the domain was used, as a means of determining daytime conditions.

Potential significant GOES data issues that may impact the error in the solar insolation product include (1) sensor degradation with time and (2) sun glint effects. The effects of the latter are small. In general, GOES satellite data are available on a continual basis with high reliability (>99%). As an example, **Figure 1** shows the daily integrated solar radiation for October 16, 2018, for Puerto Rico, the USVI, Hispaniola, Jamaica, and Cuba.

3. The GOESWEB modeling framework

GOESWEB performs daily water and energy balance calculations for the island of Puerto Rico. Twenty-seven hydro-agro-climate variables are available to the public for download (**Table 1**). Downloadable formats are available as images (jpg) or in comma-separated values (csv) and Matlab® formats. The variables in **Table 1** are also available as monthly and annual averages or totals. Simplified versions of the algorithm have been developed for estimating reference ET on the islands of the USVI, Hispaniola, Jamaica, and Cuba.

ET_o is estimated by three methods: Penman-Monteith [32, 33], Priestley-Taylor [34], and Hargreaves-Samani [35]. In [36], they described the methodology used to estimate ET_o in the earliest version of the algorithm. T_{avg}, T_{min}, and T_{max} values were estimated from a lapse rate method developed by [37]. T_d was assumed to be equal to the minimum daily T_{min} [38]. Wind speed was assumed to be the world-wide average 2-m wind speed of 2 $m(s)^{-1}$ [32]. The algorithms for Hispaniola, Jamaica, Cuba, and the USVI continue to use these simplified methods for estimating daily values of ET_o.

Water and energy balances were added to the algorithm for Puerto Rico. The daily meteorological data used are described below. **Table 2** summarizes the GOESWEB input data sources.

- Solar radiation

    i. Solar radiation (R_s) is derived from the GOES satellite using the methodology described above.

    ii. The ground level, 1-km resolution R_s product became available in Puerto Rico in March of 2009 and has been validated at two locations in Puerto Rico by [39].

    iii. Occasionally the satellite-derived solar radiation is not available, in which case the previous days' R_s values are used.

    iv. Prior to GOES–16, 1 km GOES–12 and –13 visible channel 1 data were used over Puerto Rico and the USVI and 2 km data over the other islands.

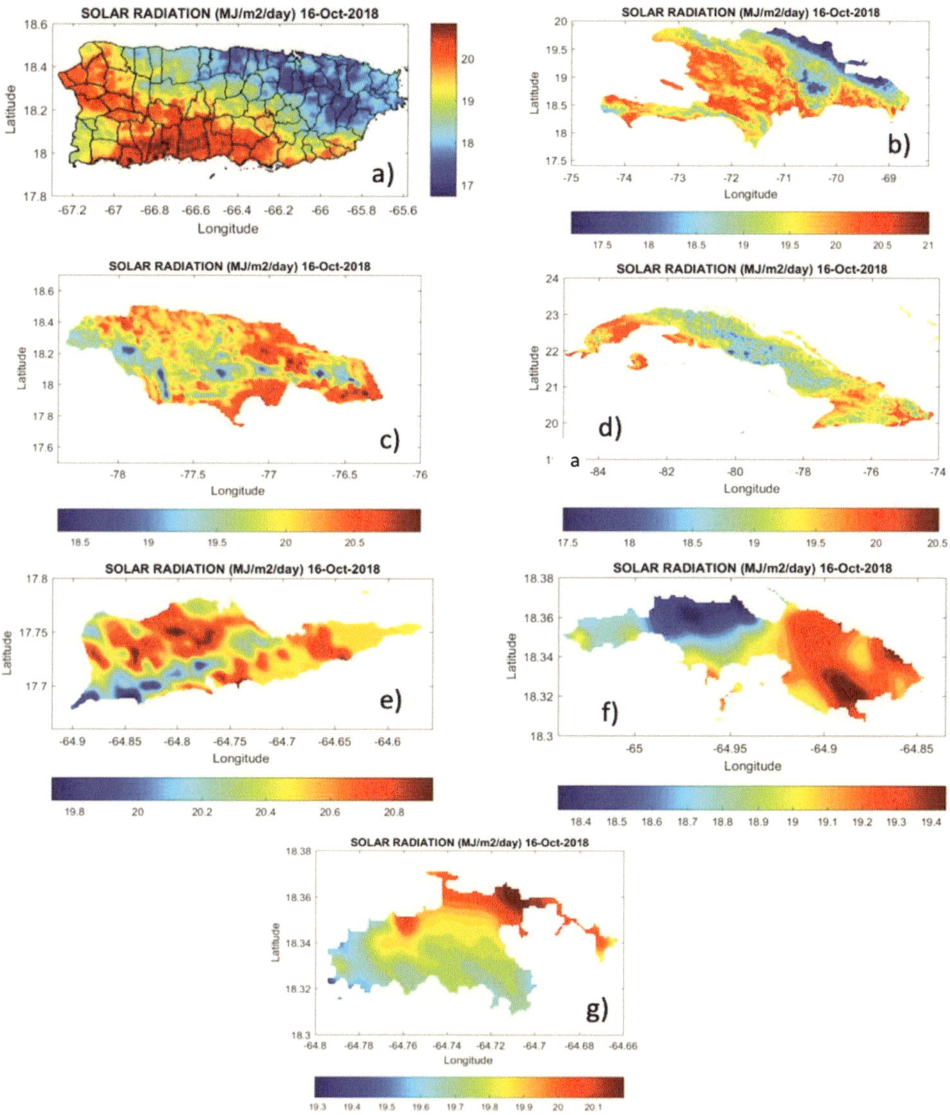

Figure 1.
Daily solar insolation for (a) Puerto Rico, (b) Hispaniola, (c) Jamaica, (d) Cuba, and (e) St. Croix USVI, (f) St. Thomas, and (g) St. John (USVI) on October 16, 2018.

1	ET	Actual evapotranspiration (mm)
2	ET_o	Reference evapotranspiration (mm); three methods
3	T_{avg}	Average air temperature (°C)
4	T_{min}	Minimum air Temperature (°C)
5	T_{max}	Maximum air Temperature (°C)
6	T_d	Dew point temperature (°C)
7	T_s	Effective surface temperature (°C)
8	e_a	Actual vapor pressure (kPa)
9	e_s	Saturated vapor pressure (kPa)
10	RH	Relative humidity (%)
11	u	Wind speed ($m\ s^{-1}$)
12	R_s	Solar radiation ($MJ\ m^{-2}\ day^{-1}$)
13	R_n	Net radiation ($MJ\ m^{-2}\ day^{-1}$)
14	PAR	Photosynthetically active radiation (micromoles $m^{-2}\ sec^{-1}$)
15	K_s	Water stress coefficient (unitless)
16	$K_{c,eff}$	Effective crop coefficient (unitless)
17	Precip	Rainfall (mm)
18	$Precip_{eff}$	Effective rainfall (mm)
19	RO	Surface runoff (mm)
20	DP	Deep percolation (mm)
21	θ	Soil moisture content ($m^{-3}\ m^{-3}$)
22	Sat	Soil moisture saturation (unitless)
23	r_s	Surface resistance ($s\ m^{-1}$)
24	r_a	Aerodynamic resistance ($s\ m^{-1}$)
25	LE	Latent heat flux ($MJ\ m^{-2}\ day^{-1}$)
26	H	Sensible heat flux $MJ\ m^{-2}$ day-1)
27	β	Bowen ratio (unitless)

Table 1.
Hydro-agro-climate variables produced daily by GOESWEB.

Model input	ET_o	ET_a	Source
Solar radiation	X	X	GOES
Air temperature	X	X	NOAA/NDFD, CariCOOS/WRF
Dew point temperature	X	X	NOAA/NDFD, CariCOOS/WRF
Wind speed (2-m height)	X	X	NOAA/NDFD, CariCOOS/WRF
Albedo[1], root depth, roughness length, zero-plane displacement		X	ATMET, 2005
Soil texture		X	USDA/SSURGO

[1]*Albedo is 0.23 for calculating ET_o.*

Table 2.
Summary table of required GOESWEB input data sources.

- Air temperatures

  i. T_{avg}, T_{min}, T_{max}, and T_d data were obtained from the National Digital Forecast Database (NDFD) website [40] from January 1, 2009, to December 31, 2016.

  ii. Temperatures were obtained from the CariCOOS operational gridded Weather Research and Forecasting (WRF) model starting on January 1, 2017.

  iii. On occasion, weather parameters from the WRF model are not available. NDFD air temperatures and wind speed are used in those cases. As a final resort, the lapse rate method of [37] is used.

- Wind speed

  i. During the period January 1, 2009, through September 30, 2015, daily average wind speed was obtained from the average of eight NDFD 3-hour values [40].

  ii. From October 1, 2015, to the present, daily average wind speed was obtained from the average of 24-hourly wind speed values obtained from the Caribbean Coastal Ocean Observing System (CariCOOS) WRF.

  iii. For the reference ET calculation, 10-m wind speeds are adjusted to 2 meters [32].

  iv. If wind speed is not available, then the previous day's data are used.

3.1 Net radiation calculations

Net radiation (R_n) is estimated using the methodology described by [32] and used by [41].

$$R_n = R_{ns} + R_{nl} \tag{1}$$

where R_n is net radiation, R_{ns} is net shortwave radiation, and R_{nl} is net long wave radiation.

$$R_{ns} = (1 - \alpha)R_s \tag{2}$$

where α is albedo and R_s solar radiation. α is defined as 0.23 for estimating RET, and values are derived from a lookup table associated with 32 land cover classes [42] for estimating actual ET (ET_a). R_s is derived from the GOES satellite. The net long wave radiation is estimated from the equation

$$R_{nl} = \sigma\left[\frac{T_{max,K}{}^4 - T_{min,K}{}^4}{2}\right](0.34 - 0.14\sqrt{e_a})\left(1.35\frac{R_s}{R_{so}} - 0.35\right) \tag{3}$$

where σ is the Stefan-Boltzmann constant, T_{max} is maximum absolute temperate during the 24-hour period, T_{min} is minimum absolute temperature during the 24-hour period, e_a is actual vapor pressure, R_s/R_{so} is relative shortwave radiation (limited to $\leq$1.0), and R_{so} calculated clear-sky radiation. Actual vapor pressure is estimated by

$$e_a = 0.6108 \ \exp \frac{(17.27T_d)}{(T_d + 237.3)} \tag{4}$$

where T_d is dew point temperature. The calculated clear-sky radiation is estimated by

$$R_{so} = \left(0.75 + 2\,10^{-5}\,z\right) R_a \tag{5}$$

where z is elevation above mean sea level and R_a is extraterrestrial radiation.

$$R_a = \frac{12(60)}{\pi} G_{sc} d_r [(\omega_2 - \omega_1) \sin(\varphi) \sin(\delta) + \cos(\varphi) \cos(\delta)(\sin(\omega_2) - \sin(\omega_1))] \tag{6}$$

where G_{sc} is the solar constant = 0.0820 and d_r is the relative distance Earth-Sun, defined as

$$d_r = 1 + 0.33 \cos\left(\frac{2\pi}{365} J\right) \tag{7}$$

where J is Julian day (e.g., January 1 is 1 and December 31 is 365). ω_1 in Eq. (6) is solar time angle at the beginning of the period and ω_2 solar time angle at end of period, generally expressed as

$$\omega_s = \frac{\pi}{2} - \arctan\left[\frac{-\tan(\varphi)\tan(\delta)}{X^{0.5}}\right] \tag{8}$$

where φ is latitude and δ solar declination expressed as

$$\delta = 0.409 \sin\left(\frac{2\pi}{365} J - 1.39\right) \tag{9}$$

and X is defined as

$$X = 1 - [\tan(\varphi)]^2 [\tan(\delta)]^2 \tag{10}$$

and X = 0.00001 if $X \leq 0$.

3.2 Reference evapotranspiration estimates

The Penman-Monteith (PM) equation is given by Eq. 1 [32], which applies specifically to a hypothetical reference crop with an assumed crop height of 0.12 m, an albedo of 0.23, a fixed surface resistance of 70 sec m^{-1}, and an aerodynamic resistance equal to $208/u_2$, where u_2 is wind speed at 2 m height:

$$ET_o = \frac{0.408\Delta(R_n - G) + \gamma\left(\frac{900}{T+273}\right) u_2 (e_s - e_a)}{\Delta + \gamma(1 + 0.34u_2)} \tag{11}$$

where Δ is the slope of the vapor pressure curve, G is soil heat flux, γ is the psychrometric constant, T is mean daily temperature at 2 m height, e_s is the saturation vapor pressure, and e_a is the actual vapor pressure.

The second method used to estimate ET_o is the Priestly-Taylor Equation [34], a simplification of the Penman Equation [43, 44]:

$$ET_o = \alpha \frac{\Delta(R_n - G)}{\Delta + \gamma} \quad (12)$$

where α is the Priestly-Taylor constant. Values in the literature for α range from 1.26 [34] to 1.32 [45]. In this study we use a value of α equal to 1.3.

The third method used to estimate ET_o is the Hargreaves-Samani ET_o Equation [35] given by

$$ET_o = 0.408[0.0135R_s](T + 17.8) \quad (13)$$

The value 0.0135 is a constant and 0.408 converts the result from MJ m^{-2} day to mm $(day)^{-1}$. In [38] they showed that this method produces comparable results with the PM method in PR.

The PM method is considered superior to the other two methods because it accounts for the major variables that control ET (R_n, T, VPD and u), and the PM method has been rigorously validated [33].

3.3 Energy balance

In GOESWEB, an energy balance approach is used similar to [46]. The basic energy balance equation is given as

$$R_n - LE - H - G = 0 \quad (14)$$

R_n is obtained from the calculation procedure presented above. Albedo, which is used in the R_n calculation, is obtained from a lookup table [42], which assigns values of the parameters to 32 different land covers.

LE, H, and G are the latent, sensible, and soil heat fluxes, respectively. LE is estimated using the following Equation [47]:

$$LE = \frac{\rho C_p(e_o(T_s) - e(T_a))}{\gamma(r_a + r_s)} \quad (15)$$

where ρ is mean air density, C_p is specific heat, r_a is aerodynamic resistance, and r_s is surface resistance. G is the soil heat flux, assumed to be zero for the daily analysis. H is estimated using the following equation:

$$H = \frac{\rho C_p(T_s - T_a)}{r_a} \quad (16)$$

The effective surface temperature is difficult to obtain from remote sensing under cloudy conditions. Therefore, T_s is obtained by an implicit approach similar to that described by [48]. When Eq. (14) is expanded using Eqs. (1), (15), and (16), T_s is the only unknown variable, which is obtained using the recursive root function **fzero** in MatLab® (http://www.mathworks.com).

The aerodynamic resistance (r_a) is calculated with the following Equation [46]:

$$r_a = r_{ao}\phi + r_{bh} \quad (17)$$

where r_{ao} is the aerodynamic resistance under conditions of neutral atmospheric stability and r_{bh} is the excessive resistance. r_{ao} is expressed as

$$r_{ao} = \frac{ln\left[\frac{z-z_{disp}}{z_o}\right] ln\left[\frac{z-z_{disp}}{(0.1)z_o}\right]}{k^2 u} \tag{18}$$

where z is the virtual height at which meteorological measurements are taken. In this study z is assumed to be within the inertial sublayer and equal to $1.5(z_o/0.13)$ [47], which is equivalent to the canopy height (h). The NDFD or WRF model-derived wind speeds at 10 m height are adjusted to the "virtual instrument height," depending on the height of the vegetation. Roughness length (z_o) and the zero-plane displacement (z_{disp}) are derived from a lookup table for various land use/vegetation categories [42]. k is Von Karman's constant (k = 0.41). u is the wind velocity at height z.

From [46], the atmospheric stability coefficient is

$$\phi = \left[1 - \frac{\left[\eta\left(z - z_{disp}\right)g(T_s - T_a)\right]}{T_o u^2}\right] \tag{19}$$

where g is the gravitational constant and the coefficient η is taken as 5 [46]. The temperature, T_o, is the average of the values of T_s and T_a. Other variables and parameters were previously defined.

The excess resistance in Eq. (17) is given by the equation

$$r_{bh} = \frac{4}{\left(\frac{ku}{ln\left[\left(z - z_{disp}\right)/z_o\right]}\right)} \tag{20}$$

Bulk surface resistance (r_s) is estimated using the equation of [49]:

$$r_s = \frac{\rho C_p VPD}{\Delta(R_n - G)C_f}\left(\frac{\theta - \theta_{WP}}{\theta_{FC} - \theta_{WP}}\right)^{-1} \tag{21}$$

where VPD is the vapor pressure deficit, C_f is a calibration coefficient equal to 1 for root depth <1 m and 5 for root depth >1 m, and θ_{FC} and θ_{WP} are the volumetric soil moisture content (θ) at field capacity and wilting point, respectively. Field capacity and wilting point were obtained from regression equations of [50] based on percent sand, silt, and clay. Soil properties for sand, silt, and clay for Puerto Rico were obtained from the USDA Natural Resource Conservation Service (NRCS) and Soil Survey Geographic (SSURGO) database.

3.4 Water balance

The water balance is estimated from the equation

$$SMD2 = Precip - ET_a - RO - DP + SMD1 \tag{22}$$

where SMD1 and SMD2 are the depths of soil moisture in the root zone (Rdepth) at times 1 and 2, respectively. In GOESWEB the time step is 1 day. Precip is rainfall, RO is surface runoff, and DP is deep percolation below the root zone. The daily ET_a is obtained by converting LE to an equivalent depth of water by dividing by the latent heat of vaporization (2.45 MJ kg^{-1}). Root depths for various land use/vegetation categories are obtained from [42] lookup table. Twenty-four-hour rainfall is obtained from NOAA's Advanced Hydrologic Prediction Service (AHPS). In PR, AHPS rainfall is bias-corrected radar rainfall using rain gauge data.

Surface runoff is estimated using the curve number (CN) method of the NRCS [51]:

$$RO = \frac{(Precip - 0.2S)^2}{(Precip + 0.8S)} \tag{23}$$

$$S = \left[\left(\frac{25400}{CN}\right) - 254\right] \tag{24}$$

where S is the maximum potential difference between rainfall and runoff at the moment of rainfall initiation and CN is a proportion of rainfall converted to runoff, adjusted for antecedent rainfall conditions. CN values were derived for Puerto Rico using the method described by [51], based on land use, hydrologic soil group, and antecedent rainfall conditions.

To estimate DP, the following procedure is followed: $SMD2_i$ = Precip – ET_a – RO + SMD1. If the value of $SMD2_i$ is larger than the depth of water in the soil profile at field capacity (FCD), then DP = $SMD2_i$ – FCD, and the value of SMD2 is equal to FCD. If $SMD2_i$ < FCD, then DP = 0 and SMD2 = $SMD2_i$.

3.5 GOESWEB model accuracy and validation

In this section accuracy and validation data are presented for remotely sensed solar radiation, RET, soil moisture, and stream flow. Solar radiation is a critically important variable in the estimation of ET. **Figures 2** and **3** show comparisons of the daily integrated solar radiation at the University of Puerto Rico (UPR) Fortuna Agricultural Experimental near Juana Diaz, PR, and the UPR-Mayaguez Campus (UPRM) in Mayaguez, PR, respectively [39]. The figures show a high degree of correlation between the remote sensing solar radiation and the measured solar radiation. The coefficients of determination (r^2) for the UPRM and experimental station data were 0.88 and 0.83, respectively. (From [39]).

Figure 4 shows a comparison of the ET_o computed by the GOESWEB algorithm and from weather station data from the UPR Fortuna Agricultural Experiment Station, near Juana Diaz, PR. The ET_o data covers the period from December 12, 2013 to April 20, 2016 (858 days). Although the vast majority of data pairs fall close to the 1:1 line, indicating close agreement between the two methods, a smaller number of data pairs fall relatively far from the 1:1 line, producing the scatter in the

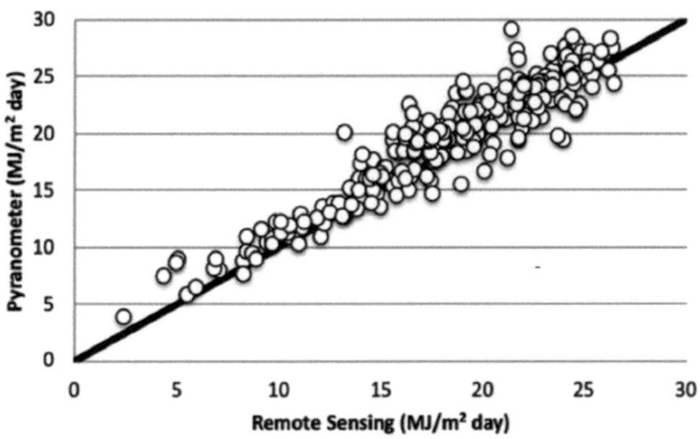

Figure 2.
Comparison of remote sensing and pyranometer-measured daily integrated solar radiation at the UPR Fortuna Agricultural Experiment Station, near Juana Diaz, PR (From [39]). The r^2 value for this comparison is 0.88.

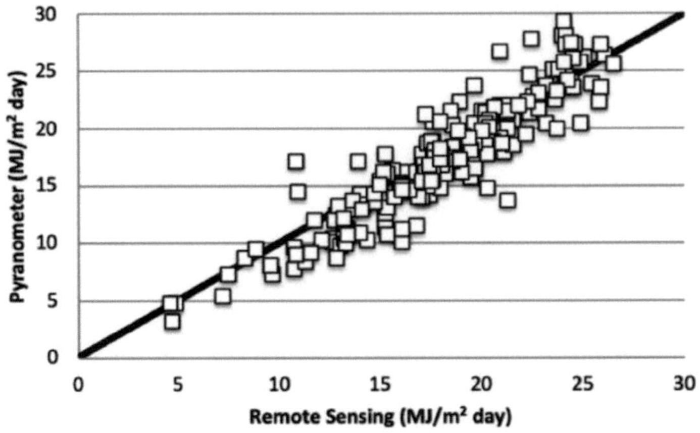

Figure 3.
Comparison of remote sensing and pyranometer-measured daily integrated solar radiation at UPRM (From [39]). The r^2 value for this comparison is 0.83.

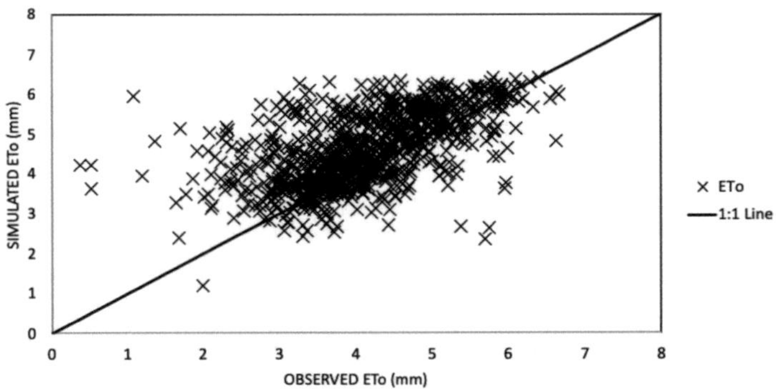

Figure 4.
Comparison of observed and simulated ET_o at the UPR Fortuna Agricultural Experiment Station, near Juana Diaz, PR. The data cover the period December 12, 2013 through April 20, 2016. The r^2 value for this comparison is 0.31.

data set. For this comparison, the coefficient of determination (r^2) was 0.31. The average GOESWEB and weather station ET_o were 4.6 mm and 4.14 mm, respectively, and the average calculated error was 11.2%. It should be noted that the weather station at this location does not comply with the required "reference conditions" for computing ET_o. Reference conditions refer to a grass-type vegetation with an approximate height of 0.12 m, an albedo of 0.23, and a fixed surface resistance of 70 sec m^{-1}, receiving adequate water. The climate of southern PR is semiarid, and there are frequent times when there was no vegetation at all on the ground surrounding the weather station.

Figure 5 shows a time series comparison of soil moisture from GOESWEB and soil moisture from a weather station located at the UPR Fortuna Agricultural Experiment Station. The weather station soil moisture is an average of five sensors positioned at depths of 0.0508 m (2 in.), 0.1016 m (4 in.), 0.2032 m (8 in.), 0.508 m (20 in.), and 1.016 m (40 in.). Immediately after rainfalls the weather station soil moisture tended to rise to higher soil moisture values than the soil moisture from the model. It is important to know that maximum soil moisture values in GOESWEB are limited to the field capacity, as excess water is routed below the root zone as deep percolation. Furthermore, the sensor soil moisture represent a single point (approximately 1 m^2), whereas the model represents an

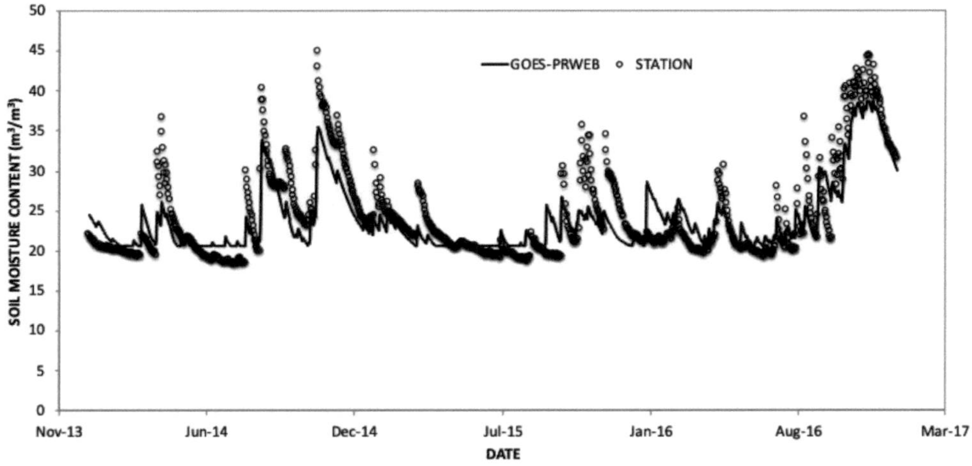

Figure 5.
Comparison of the GOESWEB and weather station soil moisture at the UPR Fortuna Agricultural Experiment Station, near Juana Diaz, PR. The data covers the period January 1, 2014–December 31, 2016. The r^2 value for this comparison is 0.73.

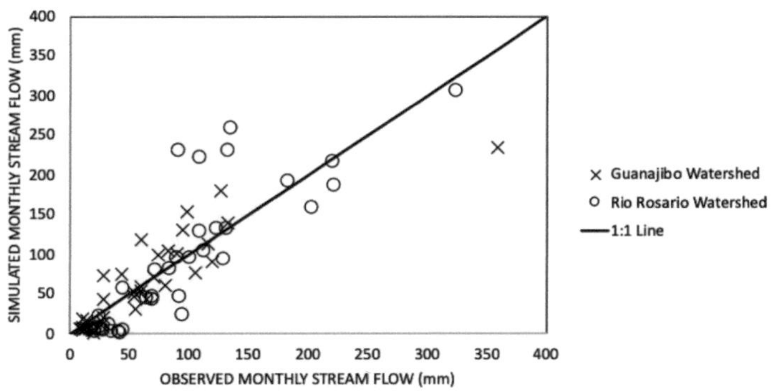

Figure 6.
Comparison of observed and simulated stream flow for two watersheds in southwest Puerto Rico. The data cover the 36-month period during 2010–2012. The r^2 value for this comparison is 0.72 for all data.

area of 1 km^2 (1,000, 000 m^2), and therefore, complete agreement between the two methods would not be expected.

Figure 6 compares the monthly stream flow values for two watersheds in southwest Puerto Rico. Observed stream flow values were obtained from the US Geological Survey (USGS). The results are presented as a depth of water in millimeters (i.e., monthly stream volume/watershed area). The total stream flow for the model was assumed to be the surface runoff plus the deep percolation (or aquifer recharge). The latter term represents the stream base flow. To obtain the monthly value of stream flow in the model, the surface runoff and deep percolation were averaged for every 1 km^2 pixel within the watershed. The model does a reasonably good job of simulating monthly stream flow.

4. High-resolution products for Puerto Rico

Figure 7 shows an example of selected water and energy balance components for Puerto Rico on October 16, 2018. Rainfall, surface runoff, percolation below the

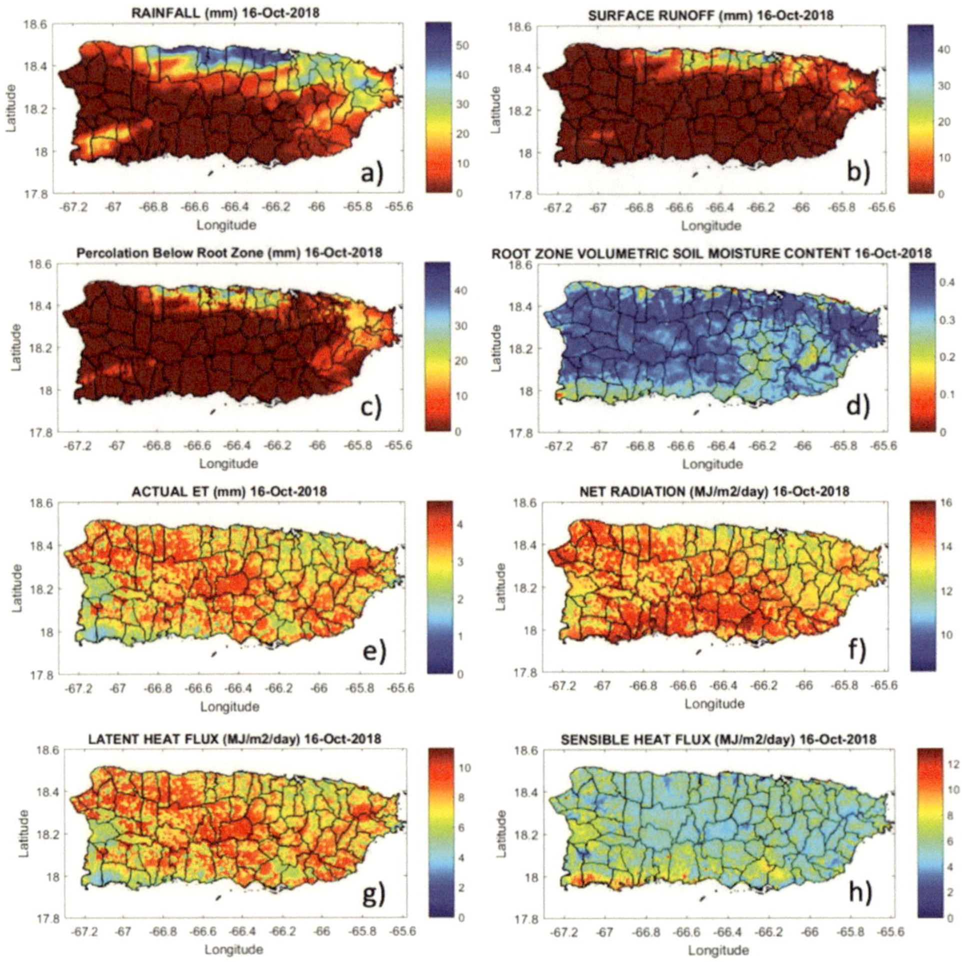

Figure 7.
Example of water and energy balance components from the GOESWEB algorithm for November 24, 2015.

Figure 8.
NOAA's Drought Monitor for Puerto Rico, October 18, 2018. The municipalities of Aibonito, Cayey, and Cidra are experiencing abnormally dry conditions.

root zone, soil moisture content, actual ET, Rn, LE, and H are included. Approximately 60 mm of rain fell along the northern coast of the island. High values of surface runoff occurred in the rainy area where soil textures have high clay content. High values of percolation below the root zone occurred in small areas where the soil sand content approaches 90%. The soil moisture map indicates a lobe of dry area in Salinas, Cayey, Aibonito, and Cidra. **Figure 8** shows the NOAA Drought Monitor for Puerto Rico for October 18, 2018, indicating abnormally dry conditions for Cayey, Aibonito, Cidra, and a portion of Barranquitas. The figure also shows LE

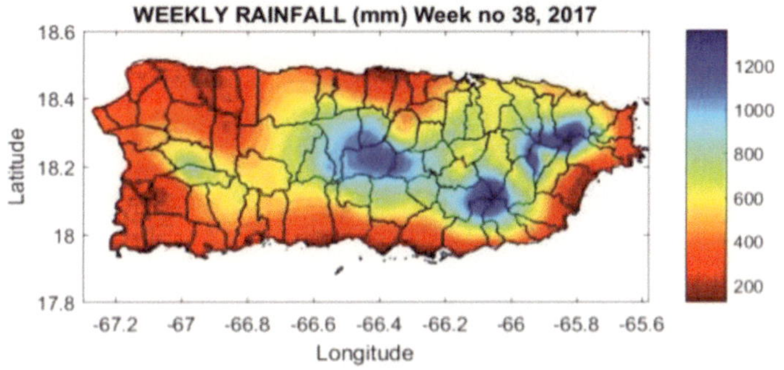

Figure 9.
Rainfall over Puerto Rico during the week of Hurricane Maria, during September 2017.

and H fluxes, which sum to the Rn (i.e., Eq. (14)). ETa is the LE flux divided the latent heat of vaporization constant equal to 2.45 MJ kg^{-1}.

Figure 9 shows the rainfall during the week of September 17, 2017, the same week Hurricane Maria occurred. The maximum rainfall for the week was nearly 1300 mm (51 in.) in southeast Puerto Rico. The rainfall data were derived from rain gauge data, since the Doppler radar in Cayey, PR, was severely damaged during the hurricane. The National Weather Service (NWS) combined the gauge rainfall for September 20 and 21. The maximum rainfall during the 2-day period was 950 mm

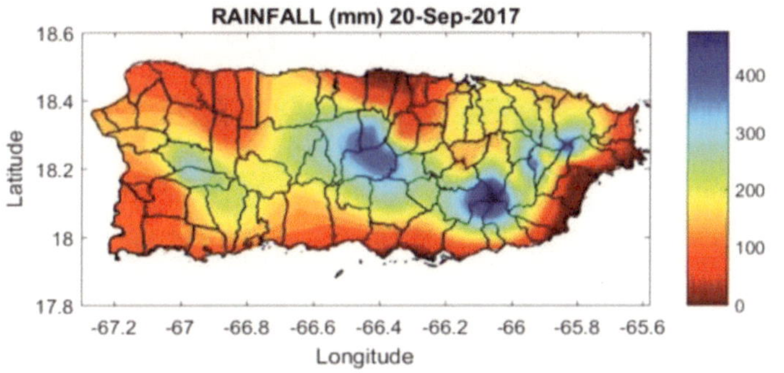

Figure 10.
Rainfall over Puerto Rico on September 20, the day that Hurricane Maria made landfall on Puerto Rico. The gauge rainfall reported by the NWS was for the 20th and 21st; therefore the rainfall for September 20 was assumed to be half the amount.

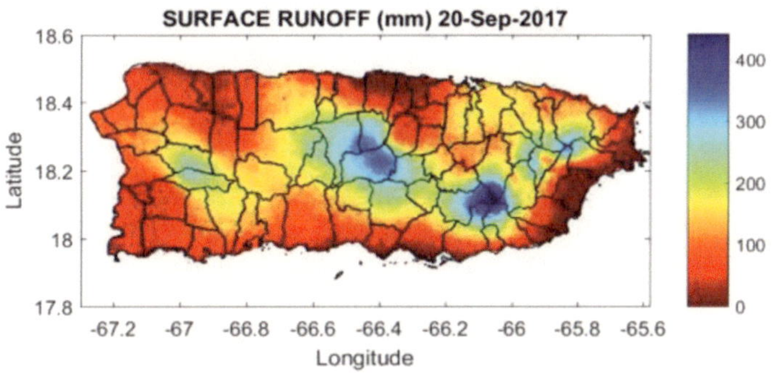

Figure 11.
Estimated surface runoff over Puerto Rico on September 20, the day that Hurricane Maria made landfall on Puerto Rico.

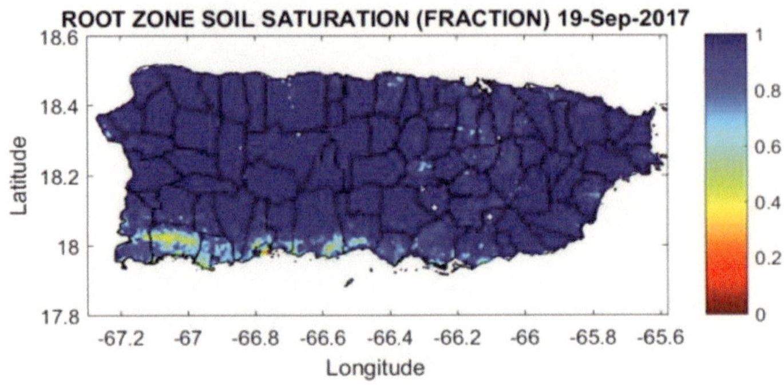

Figure 12.
Root zone soil moisture saturation for September 19, 2018, 1 day before Hurricane Maria made landfall on Puerto Rico.

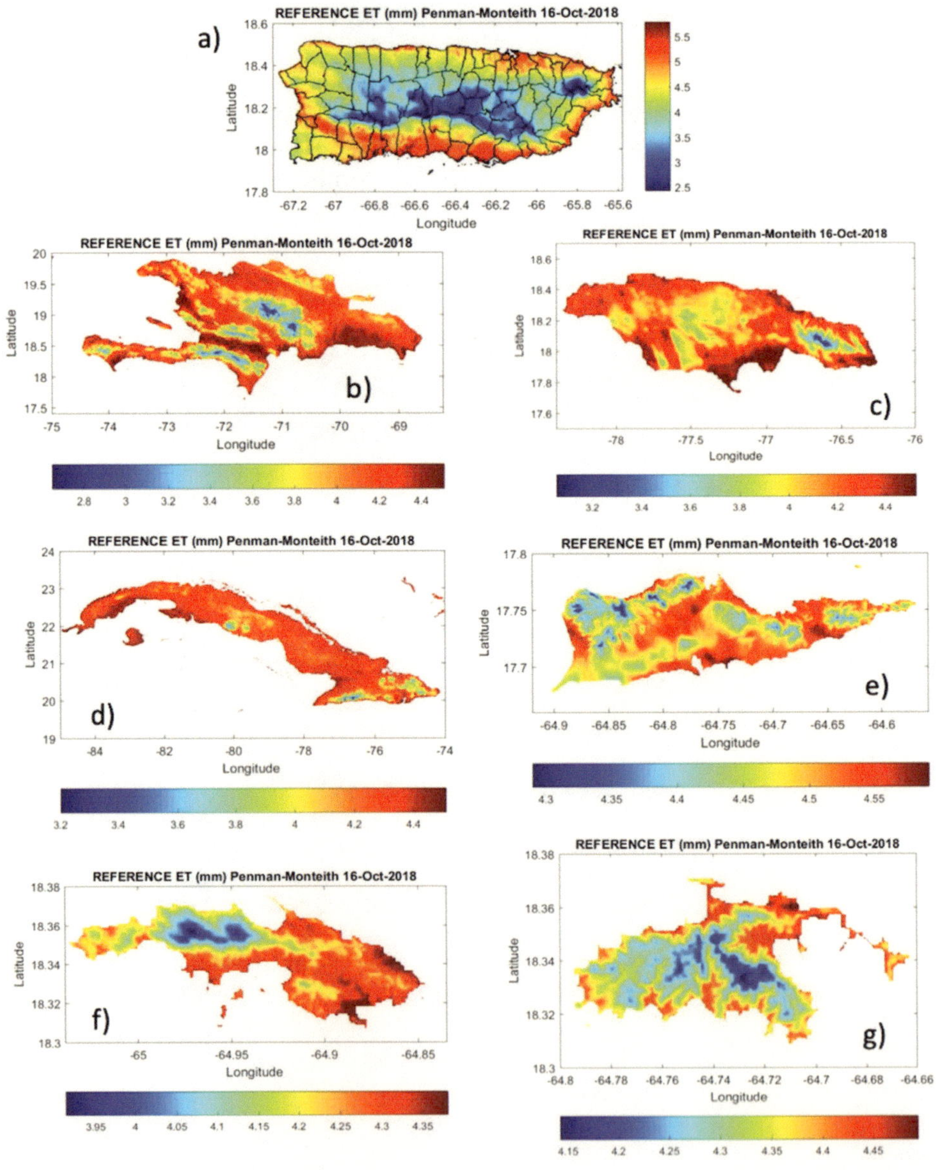

Figure 13.
ET_o for October 16, 2018, for (a) Puerto Rico, (b) Hispaniola, (c) Jamaica, (d) Cuba, and (e) St. Croix, (f) St. Thomas, and (g) St. John (USVI).

(37.5 in.) in southeast Puerto Rico. To simulate the daily hydrology, rainfall was evenly divided between the 2 days. **Figures 9** and **10** show the rainfall and surface runoff for September 20, respectively. Note that the surface runoff is almost identical to the rainfall, as seen in **Figure 11**. Nearly 100% of the rainfall was converted to surface runoff because the soils were already saturated the day before Hurricane Maria arrived (September 19), as shown in **Figure 12**.

5. High-resolution ETo products across the Caribbean

GOESWEB provides daily values of ET_0 for Puerto Rico, the USVI, Hispaniola, Jamaica, and Cuba. As an example, the ET_0 for each of the islands for October 16, 2018, are presented in **Figure 13**. In the study by [52], they describe a web-based method for determining irrigation requirements using the GOESWEB ET_0 maps.

6. Conclusions

The above study demonstrates the operational utility of incorporating spatially continuous, high spatial resolution (1 km) GOES–16-derived solar insolation, using the model described by [24], into the water balance model GOESWEB, to then estimate the complete water budget. In this demonstration, applications of water balance were performed over the US territory of Puerto Rico, a subtropical location that is very sensitive to high rates of ET, relative to various crop types and vegetation characteristics, and that also receives high amounts of rainfall. High rainfall causes significant runoff, for which the GOESWEB water balance model can help identify related to actual rainfall events. Expanding GOESWEB to other island regions would be a future avenue for the research and algorithm development activities described here.

Acknowledgements

This work was supported by the NOAA-CREST (grant NA06OAR4810162), National Science Foundation (grant 0313747 and 1832576), and the U.S. Department of Agriculture (Hatch Project H402). Any opinions, findings, conclusions, or recommendations expressed in this material are those of the authors and do not necessarily reflect those of the National Oceanic and Atmospheric Administration, NSF, or the USDA. Reference to a commercial product in no way constitutes an endorsement of the product by the authors.

Conflict of interest

None.

Notes/thanks/other declarations

Thanks to Luis Aponte-Bermudez for providing the Matlab® computer code and for reading the daily WRF wind speed data sets and CariCOOS for providing the WRF wind speed, air temperature, and relative humidity data. There are numerous other people that directly or indirectly assisted in this research and to those people we would like to express our appreciation.

Nomenclature

B (unitless)	Bowen ratio
C_f (dimensionless)	calibration coefficient
CN (dimensionless)	curve number, proportion of rainfall converted to runoff
DP (mm)	deep percolation or the soil water that passes below the root zone
d_r (dimensionless)	relative distance Earth-Sun
ET_a (mm)	actual evapotranspiration
ET_o (mm)	reference evapotranspiration
e_a [kPa]	actual vapor pressure
e_s (kPa)	saturated vapor pressure
g ($m\ s^{-2}$)	gravitational constant
G ($MJ\ m^{-2}\ day^{-1}$)	soil heat flux density
G_{sc} ($MJ\ m^{-2}\ min^{-1}$)	solar constant = 0.0820
h (m)	canopy height
H ($MJ\ m^{-2}\ day^{-1}$)	sensible heat flux
J	Julian day (e.g., January 1 is 1 and December 31 is 365)
k (dimensionless)	Von Karman's constant (0.41)
$K_{c,eff}$ (unitless)	effective crop coefficient
K_s (unitless)	water stress coefficient
LE ($MJ\ m^{-2}\ day^{-1}$)	latent heat flux
PAR (micromoles $m^{-2}\ sec^{-1}$)	photosynthetically active radiation
Precip (mm)	rainfall precipitation
$Precip_{eff}$ (mm)	effective rainfall
R_a ($MJ\ m^{-2}\ day^{-1}$)	extraterrestrial radiation
r_a ($s\ m^{-1}$)	aerodynamic resistance
r_{ao} ($s\ m^{-1}$)	aerodynamic resistance under conditions of neutral atmospheric stability
r_{bh} ($s\ m^{-1}$)	excess resistance
R_{depth} (mm)	root depth
RH (%)	relative humidity
R_n ($MJ\ m^{-2}\ day^{-1}$)	net radiation
R_{nl} ($MJ\ m^{-2}\ day^{-1}$)	net long wave radiation
R_{ns} ($MJ\ m^{-2}\ day^{-1}$)	net shortwave radiation
RO (mm)	surface runoff
R_s ($MJ\ m^{-2}\ day^{-1}$)	solar radiation
r_s ($s\ m^{-1}$)	surface resistance
R_s/R_{so}	relative shortwave radiation
R_{so} ($MJ\ m^{-2}\ day^{-1}$)	clear-sky radiation
S (mm)	maximum potential difference between rainfall and runoff at the moment of rainfall initiation
Sat (unitless)	soil moisture saturation
SMD1 (mm)	depths of water in the soil profile beginning of the day (24 hours)
SMD2 (mm)	depths of water in the soil profile end of the day (24 hours)
T_a (°C)	air temperature
T_{avg} [°C]	mean daily air temperature
T_d (°C)	dew point temperature

T_{max} (K)	maximum absolute temperate during the 24-hour period
T_{max} (°C)	maximum air temperature
T_{min} (°C)	minimum air temperature
T_{min} (K)	minimum absolute temperature during the 24-hour period
T_o (°C)	average of the values of T_s and T_{avg}
T_s (°C)	effective surface temperature
U (m s^{-1})	wind speed
u_2 (m s^{-1})	wind speed at 2 m height
VPD (kPa)	vapor pressure deficit
z (m)	elevation above mean sea level
z_{disp} (m)	zero-plane displacement
z_o (m)	roughness length
α	Albedo and Priestly-Taylor constant
δ (rad)	solar declination
Δ [kPa $°C^{-1}$]	slope of the vapor pressure curve
η (dimensionless)	coefficient, commonly taken as 5
φ (rad)	latitude
σ (MJ K^{-4} m^{-2} day^{-1})	Stefan-Boltzmann constant (4.903 10^{-9} MJ K^{-4} m^{-2} day^{-1})
γ (kPa $°C^{-1}$)	psychrometric constant
ω_1 (rad)	solar time angle at beginning of period
ω_2 (rad)	solar time angle at end of period
θ (m^{-3} m^{-3})	soil moisture content
θ_{FC} (m^3/m^3)	θ values at field capacity
θ_{WP} (m^3/m^3)	θ values at wilting point

Author details

John R. Mecikalski[1]* and Eric W. Harmsen[2]

1 University of Alabama in Huntsville, Huntsville, Alabama, United States of America

2 University of Puerto Rico–Mayaguez Campus, Mayaguez, Puerto Rico, United States of America

*Address all correspondence to: johnm@nsstc.uah.edu

References

[1] Tarpley JD, Schneider SR, Bragg JE, Waters MP III. Satellite data set for solar incoming radiation studies. NOAA Technical Memorandum. 1978;**96**. 36 pp

[2] Diak GR, Gautier C. Improvements to a simple physical model for estimating insolation from GOES data. Journal of Climate and Applied Meteorology. 1983;**22**:505-508

[3] Tarpley JD. Estimating incident solar radiation at the surface from geostationary satellite data. Journal of Applied Meteorology. 1978;**18**:1172-1181

[4] Gautier C, Diak GR, Masse S. A simple physical model to estimate incident solar radiation at the surface from GOES satellite data. Journal of Applied Meteorology. 1980;**19**: 1007-1012

[5] Gautier C, Diak GR, Masse S. An investigation of the effects of spatially averaging satellite brightness measurements on the calculation of insolation. Journal of Climate and Applied Meteorology. 1984;**23**: 1380-1386

[6] Möser W, Raschke E. Incident solar radiation over Europe from METEOSAT data. Journal of Climate and Applied Meteorology. 1984;**23**:166-170

[7] Pinker RT, Ewing JA. Modeling surface solar radiation: Model formulation and validation. Journal of Climate and Applied Meteorology. 1985; **24**:389-401

[8] Dedieu G, Deschamps PY, Kerr YH. Satellite estimates of solar irradiance at the surface of the earth and of surface albedo using a physical model applied to meteosat data. Journal of Climate and Applied Meteorology. 1987;**26**:79-87

[9] Darnell WL, Staylor WF, Gupta SK, Denn FM. Estimation of surface insolation using sun-synchronous satellite data. Journal of Climate. 1988;**1**: 820-835

[10] Frouin R, Chertock B. A technique for global monitoring of net solar irradiance at the ocean surface. Part I: Model. Journal of Applied Meteorology. 1992;**31**:1056-1066

[11] Pinker RT, Laszlo I. Modeling surface solar irradiance for satellite applications on global scale. Journal of Applied Meteorology. 1992;**31**:194-211

[12] Weymouth G, LeMarshall L. An operational system to estimate global solar exposure over the Australian region from satellite observations. Australian Meteorological Magazine. 1999;**48**:181-195

[13] Schmetz J. Towards a surface radiation climatology. Retrieval of downward irradiance from satellites. Atmospheric Research. 1989;**23**:287-321

[14] Pinker RT, Frouin R, Li Z. A review of satellite methods to derive surface shortwave irradiance. Remote Sensing of Environment. 1995;**51**:105-124

[15] Mecikalski JR, Shoemaker WB, Wu Q, Holmes MA, Paech SJ, Sumner DM. A 20-Year high-resolution GOES insolation–evapotranspiration dataset for water resource management over the State of Florida. Journal of Irrigation and Drainage Engineering. 2018;**144**(9): 04018025

[16] Stewart JB, Watts CJ, Rodriguez JC, De Bruin HAR, van den Berg AR, Garatuza-Payan J. Use of satellite data to estimate radiation and evaporation for northwest Mexico. Agricultural Water Management. 1999;**38**:181-193

[17] Otkin J, Anderson MC, Mecikalski JR, Diak GR. Validation of GOES-Based

insolation estimates using data from the United States Climate Reference Network. Journal of Hydrometeorology. 2005;**6**:460-475. https://doi.org/10.1175/JHM440.1

[18] Frouin R, Gautier C, Katsaros KB, Lind J. A comparison of satellite and empirical formula techniques for estimating insolation over the oceans. Journal of Applied Meteorology. 1988; **27**:1016-1023

[19] Cosgrove BA, Lohmann D, Mitchell KE, Houser PR, Wood EF, Schaake J, et al. Realtime and retrospective forcing in the North American Land Data Assimilation Systems (NLDAS) project. Journal of Geophysical Research. 2003; **108**(D22):8842. DOI: 10.1029/2002JD003118

[20] Cosgrove BA, Lohmann D, Mitchell KE, Houser PR, Wood EF, Schaake JC, et al. Land surface model spinup behavior in the North American Land Data Assimilation System (NLDAS). Journal of Geophysical Research. 2003; **108**(D22):8845. DOI: 10.1029/2002JD003119

[21] Paech SJ, Mecikalski JR, Sumner DM, Pathak CS, Wu Q, Islam S, et al. Satellite-based solar radiation in support of potential and reference evapotranspiration estimates over Florida: A 10-year climatology. Water Resources Research. 2009;**45**(6):1328-1342

[22] Meng C, Pinker RT, Tarpley DJ, Laszlo I. A satellite approach for estimating regional land surface energy budget for GCIP/GAPP. Journal of Geophysical Research. 2003;**108**(D22):8861. DOI: 10.1029/2002JD003088

[23] Diak GR, Bland WL, Mecikalski JR. A note on first estimates of surface insolation from GOES-8 visible satellite data. Agricultural and Forest Meteorology. 1996;**82**:219-226

[24] Diak GR. Investigations of improvements to an operational GOES-satellite-data-based insolation system using pyranometer data from the U.S. Climate Reference Network (USCRN). Remote Sensing of Environment. 2017;**195**:79-95. DOI: 10.1016/j.rse.2017.04.002

[25] Jacobs JM, Myers DA, Anderson MC, Diak GR. GOES surface insolation to estimate wetlands evapotranspiration. Journal of Hydrology. 2002;**56**:53-65

[26] Jacobs JM, Anderson MC, Friess LC, Diak GR. Solar radiation, longwave radiation and emergent wetland evapotranspiration estimates from satellite data in Florida, USA. Hydrological Sciences Journal. 2004;**49**: 461-476

[27] Mecikalski JM, Diak GR, Anderson MC, Norman JM. Estimating fluxes on continental scales using remotely sensed data in an atmosphere–land exchange model. Journal of Applied Meteorology. 1999;**38**:1352-1369

[28] Anderson MC, Kustas WP, Norman JM. Upscaling and downscaling—A regional view of the soil-plant-atmosphere continuum. Agronomy Journal. 2003;**95**:1408-1432

[29] Anderson MC, Norman JM, Mecikalski JR, Torn RD, Kustas WP, Basara JB. A multiscale remote sensing model for disaggregating regional fluxes to micrometeorological scales. Journal of Hydrometeorology. 2004;**5**:343-363

[30] Diak GR, Anderson MC, Bland WL, Norman JM, Mecikalski JR, Aune RM. Agricultural management decisions aids driven by real-time satellite data. Bulletin of the American Meteorological Society. 1998;**79**:1345-1355

[31] Anderson MC, Bland WL, Norman JM, Diak GR. Canopy wetness and humidity prediction using satellite and synoptic-scale meteorological

observations. Plant Disease. 2001;**85**: 1018-1026

[32] Allen RG, Periera LS, Raes D, Smith M. Crop evapotranspiration: Guidelines for computing crop requirements. Irrigation and Drainage. 1998. Paper No. 56. Rome, Italy: FAO

[33] Allen RG, Walter IA, Elliott RL, Howell TA, Itenfisu D, Jensen ME, et al. The ASCE Standardized Reference Evapotranspiration Equation. Reston, Virginia: American Society of Civil Engineers; 2005

[34] Priestley CHB, Taylor RJ. On the assessment of surface heat flux and evaporation using large-scale parameters. Monthly Weather Review. 1972;**100**:81-92

[35] Hargreaves GH, Samani ZA. Reference crop evapotranspiration from temperature. Applied Engineering in Agriculture - ASAE. 1985;**1**(2):96-99

[36] Harmsen EW, Mecikalski JR, Cardona-Soto MJ, Gonzalez AR, Vasquez R. Estimating daily evapotranspiration in Puerto Rico using satellite remote sensing. WSEAS Transactions on Environment and Development. 2009;**6**(5):456-465

[37] Goyal MR, González EA, Chao de Báez C. Temperature versus elevation relationships for Puerto Rico. The Journal of Agriculture of the University of Puerto Rico. 1988;**72**(3):449-467

[38] Harmsen EW, Goyal MR, Justiniano ST. Estimating evapotranspiration in Puerto Rico. Journal of Agriculture of the University of Puerto Rico. 2002;**86** (1–2):35-54

[39] Harmsen EW, Tosado P, Mecikalski JR. Calibration of selected pyranometers and satellite derived solar radiation in Puerto Rico. International Journal of Renewable Energy Technology Research. 2014;**5**(1):43-54

[40] NDFD. National Weather Service National Digital Forecast Database. 2018. Available from: http://www.weather.gov/forecasts/graphical/sectors/puertorico.php

[41] ASCE. The ASCE Standardized Reference Evapotranspiration Equation. American Society of Civil Engineers; 2005

[42] ATMET. ATMET Technical Note, Number 1, Modifications for the Transition From LEAF-2 to LEAF-3. 2005. Available from: http://www.atmet.com/html/docs/rams/RT1-leaf2-3.pdf

[43] Penman HL. Natural evaporation from open water, bare soil and grass. Proceedings. Royal Society of London. 1948;**A**(194):120-145

[44] Penman HL. Vegetation and Hydrology. Technical Communication. 1963;**53**. Commonwealth Bureau of soils, Harpenden, England

[45] Morton FI. Operational estimates of areal evapotranspiration and their significance to the science and practice of hydrology. Journal of Hydrology. 1983;**66**:1-76

[46] Yunhao C, Xiaobing L, Peijun S. Estimation of regional evapotranspiration over Northwest China by using remotely sensed data. Journal of Geographical Sciences. 2001; **11**(2):1997-1999

[47] Monteith JL, Unsworth MH. Principles of Environmental Physics. 3rd ed. Boston, MA: Academic Press; 2007. 440p

[48] Lascano RJ, van Bavel CHM. Explicit and recursive calculation of potential and actual evapotranspiration. Agronomy Journal. 2007;**99**:585-590

[49] Ortega-Farias S, Poblete C, Zuñiga M. Evaluation of a two-layer model to estimate vine transpiration and soil

evaporation for vineyards. In: Proceedings of the ASCE World Environmental and Water Resources Congress, 2008. Ahupua'a

[50] Cemek B, Meral R, Apan M, Merdun H. Pedotransfer functions for the estimation of field capacity and permanent wilting point. Pakistan Journal of Biological Sciences. 2004; **7**(4):535-541

[51] Huffman RL, Fangmeier DD, Elliot W, Workman SR. Soil and Water Conservation Engineering. 7th ed. Published by ASABE, St. Joseph, MI; 2013

[52] Harmsen EW. TECHNICAL NOTE: A Simple Web-Based Method for Scheduling Irrigation in Puerto Rico. The Journal of Agriculture of the University of Puerto Rico. 2012;**96**:3-4